576.6

Mary R. Lunt

1976

# List of titles

## *Already published*

| | |
|---|---|
| Cell Differentiation | J. M. Ashworth |
| Biochemical Genetics | R. A. Woods |
| Functions of Biological Membranes | M. Davies |
| Cellular Development | D. Garrod |
| Brain Biochemistry | H. S. Bachelard |
| Immunochemistry | M. W. Steward |
| The Selectivity of Drugs | A. Albert |
| Biomechanics | R. McN. Alexander |
| Molecular Virology | T. H. Pennington, D. A. Ritchie |
| Hormone Action | A. Malkinson |
| Cellular Recognition | M. F. Greaves |
| Cytogenetics of Man and other Animals | A. McDermott |
| RNA Biosynthesis | R. H. Burdon |
| Protein Biosynthesis | A. E. Smith |

## *In preparation*

| | |
|---|---|
| The Cell Cycle | S. Shall |
| Biological Energy Transduction | C. Jones |
| Control of Enzyme Activity | P. Cohen |
| Metabolic Control | R. Denton, C. I. Pogson |
| Polysaccharides | D. A. Rees |
| Microbial Metabolism | H. Dalton |
| Microbial Taxonomy | D. Jones |
| Molecular Evolution | W. Fitch |
| A Biochemical Approach to Nutrition | R. A. Freedland |
| Metal Ions in Biology | P. M. Harrison, R. Hoare |
| Nitrogen Metabolism in Plants and Microorganisms | A. P. Sims |
| Photobiology | J. A. Raven |
| Muscle | R. M. Simmons |
| Xenobiotics | D. V. Parke |
| Plant Cytogenetics | D. M. Moore |
| Human Genetics | J. H. Edwards |
| Population Genetics | L. M. Cook |
| Membrane Biogenesis | J. Haslam |
| Biochemical Systematics | J. H. Harborne |
| Biochemical Pharmacology | B. A. Callingham |
| Insect Biochemistry | H. H. Rees |

# OUTLINE STUDIES IN BIOLOGY

## Editor's Foreword

The student of biological science in his final years as an undergraduate and his first years as a graduate is expected to gain some familiarity with current research at the frontiers of his discipline. New research work is published in a perplexing diversity of publications and is inevitably concerned with the minutiae of the subject. The sheer number of research journals and papers also causes confusion and difficulties of assimilation. Review articles usually presuppose a background knowledge of the field and are inevitably rather restricted in scope. There is thus a need for short but authoritative introductions to those areas of modern biological research which are either dealt with in standard introductory textbooks or are not dealt with in sufficient detail to enable the student to go on from them to read scholarly reviews with profit. This series of books is designed to satisfy this need.

The authors have been asked to produce a brief outline of their subject assuming that their readers will have read and remembered much of a standard introductory textbook of biology. This outline then sets out to provide by building on this basis, the conceptual framework within which modern research work is progressing and aims to give the reader an indication of the problems, both conceptual and practical, which must be overcome if progress is to be maintained. We hope that students will go on to read the more detailed reviews and articles to which reference is made with a greater insight and understanding of how they fit into the overall scheme of modern research effort and may thus be helped to choose where to make their own contribution to this effort.

These books are guidebooks, not textbooks. Modern research pays scant regard for the academic divisions into which biological teaching and introductory textbooks must, to a certain extent, be divided. We have thus concentrated in this series on providing guides to those areas which fall between, or which involve, several different academic disciplines. It is here that the gap between the textbook and the research paper is widest and where the need for guidance is greatest. In so doing we hope to have extended or supplemented but not supplanted main texts, and to have given students assistance in seeing how modern biological research is progressing, while at the same time providing a foundation for self help in the achievement of successful examination results.

J. M. Ashworth, Professor of Biology, University of Essex.

# Molecular Virology

## T. H. Pennington and D. A. Ritchie

*Senior Lecturers at the Institute of Virology,*
*University of Glasgow*

LONDON

## CHAPMAN AND HALL

A Halsted Press Book
JOHN WILEY & SONS, INC., NEW YORK

*First published in 1975*
*by Chapman and Hall Ltd*
*11 New Fetter Lane, London EC4P 4EE*
*© 1975 T. H. Pennington and R. A. Ritchie*
*Typeset by Preface Ltd, Salisbury, Wilts and*
*printed in Great Britain by William Clowes & Sons Ltd,*
*London, Colchester and Beccles*

ISBN 0 412 13910 3

Distributed in the U.S.A.
by Halsted Press, a Division
of John Wiley & Sons, Inc. New York

**Library of Congress Cataloging in Publication Data**
Pennington, T        H
    Molecular virology.

    (Outline studies in biology)
    Bibliography: p.
    Includes index.
    1. Virology. I. Ritchie, R. A., joint author.
II. Title. [DNLM: 1. Viruses. 2. Molecular
biology. Qw160 P414m]
Qr360.P46        576'.64        75-22112
ISBN 0-470-67935-2 (Halsted)

# Contents

# 1 Introduction

**1.1 Historical development of molecular virology**
Viruses have occupied a central position in molecular biology ever since its development as an independent discipline. Indeed, molecular biology itself largely developed out of the pioneer studies of Delbrück, Luria and Hershey, who realized, in the late 1930's, that bacterial viruses (bacteriophages, often abbreviated to phages) had properties which made them uniquely suitable as a model system for an attack on one of the then outstanding problems of biology, the definition of the gene in physical and chemical terms. The favourable properties of these viruses include the rapidity of their growth, their ease of assay, and the availability of easily scored genetic markers. Taken together, this means that quantitative genetic experiments can be done very quickly. The small size of viruses also suggested that their structure would be simple, which, in turn, led to the belief that this kind of system would be suitable for biochemical and physiological studies.

During the next two decades a small group of phage workers uncovered a series of fundamental principles which laid the foundations of modern virology and which had far-reaching effects on biological research in general. These studies established the basic pattern of virus replication, confirmed the identification of the nucleic acid molecule as the genetic material, and led to fundamental advances in the understanding of gene structure. An important factor during this period was the concentration of effort on a limited number of phages, notably the *Escherichia coli* phages T2 and T4. At the same time Lwoff and his colleagues were studying phage λ, a phage of *E. coli*, work which was to lead to equally fundamental observations on the regulation of macromolecular synthesis.

The study of animal and plant viruses has its origins in the latter half of the 19th century and was largely initiated by workers in medical, veterinary and agricultural disciplines. Many of their practical successes owe little to molecular biology, stemming instead from those approaches successful in combating other parasites, such as vector control and the breeding of resistant varieties of plants. The introduction of new tissue culture techniques in the early 1950's was, however, an event crucial in the development of animal virology, both as an applied subject and as a branch of molecular biology. The development by Dulbecco of a simplified assay for the titration of animal viruses based on the standard plaque assay method used with phage was another crucial event at this time.

The development of animal virology owes much to concepts derived from phage studies and, to a lesser extent, plant viruses. However, in recent years many important features of animal viruses have been described which have no bacteriophage parallels. These include virion transcriptases, the phenomenon of virion maturation by budding from cell membranes, and cell transformation.

## 1.2 What is a virus?

A recent definition of the term virus (Luria and Darnell, 1968) runs 'Viruses are entities whose genome is an element of nucleic acid, either DNA or RNA, which reproduces inside living cells and uses their synthetic machinery to direct the synthesis of specialized particles, the virions, which contain the viral genome and transfer it to other cells'.

Viruses differ from other obligate intracellular parasites (such as Rickettsiae, and the Psittacosis group of organisms) in several fundamental respects, including:

(1) Virus particles (virions) contain only one type of nucleic acid: this can be either DNA or RNA.

(2) Virus-specified proteins are synthesized using host ribosomes.

(3) Viruses multiply by independent synthesis of their constituent parts which are then assembled to reconstitute new virus particles, rather than by growth and division.

Virus particles consist essentially of nucleic acid (the virus genome) surrounded by a protein coat. The function of the coat is to protect the nucleic acid from the harsh extracellular environment, to facilitate its entry into host cells, and, in many animal viruses, to play an important role in the initiation of virus macromolecular synthesis during the early part of infection. The structure of virus particles varies enormously in complexity. Many plant viruses, for example, contain a single small RNA molecule packaged in a coat made up of many identical copies of a single protein, whereas some animal virus particles have coats made up of multiple copies of at least 30 different proteins surrounding an extremely long DNA molecule. Likewise the size of virus particles also varies considerably from one type to another. A starting point in most schemes of virus classification is the chemical nature of the virus genome; thus viruses are grouped into those with DNA genomes and those with RNA genomes, and these groups are further subdivided into viruses with single- and double-stranded genomes.

Three major groups of viruses can be distinguished on the basis of host specificity; viruses of bacteria and blue-green algae, plant viruses, and animal viruses. Many members of these groups show specialized features connected with the problem of gaining entry into and replicating in their particular host cells. A good example of features restricted to one of these groups is the complex structure that many bacteriophages have evolved to overcome the problem of introducing their nucleic acids into their hosts through the barrier of the tough, rigid, bacterial cell wall. Animal viruses do not encounter such a barrier and plant viruses enter cells in different ways.

## 1.3 Titration of viruses

Quantitative analysis demands methods for the determination of the numbers of virus particles in a sample. Many methods are available. Some of these assay virus particles directly e.g. electron microscopy and haemagglutination of red blood cells, whereas others measure the infectious titre of a virus stock. These two types of method may not give the same titre since not all virions may be infectious. For phages there is generally a one to one correspondence between physical and infectious particles; for animal viruses the infectious titre is usually lower than the number of particles. A typical phage assay involves mixing a suitable dilution of virus (e.g. containing 100–200 infectious particles) with a concentrated suspension of bacteria (about $10^8$ cells) suspended in molten agar held at $45°C$. This mixture is poured over the surface of a petri plate containing solidified nutrient agar, to form a thin layer which soon hardens, thus immobilizing the phage and bacteria. The plate is incubated to permit multiplication of the bacteria which forms a confluent film of cells over the agar surface, except where an infectious phage particle has been deposited. At this site the virus infects a cell and multiplies within it. The crop of a hundred or so progeny liberated from this cell infect adjacent bacteria which in turn produce further virus. Thus a local chain-reaction develops which after a few hours is visible as a clearing in the otherwise dense lawn of confluent bacteria. This clear zone is known as a plaque. By counting the number of plaques a direct estimate of the number of infectious virus particles is obtained. The plaque assay for animal viruses developed by Dulbecco is basically similar. Virus is added to a sheet of tissue culture cells growing in liquid medium on the flat bottom of a dish. After allowing the virus to adsorb to the cells, a layer of molten agar containing nutrient medium is poured over the cells and allowed to harden. This prevents free diffusion of virus through the medium. As with the phage assay, local areas of virus growth, each starting from one infected cell, are produced in the cell sheet. As cell death rather than cell lysis is the common end result of animal virus infections, plaques are usually detected by staining with dyes which are only taken up by living cells; the plaques stand out as colourless areas against a stained background of uninfected cells.

## 1.4 Virus–host interactions

Cell death is a common end result of virus infection. Many bacteriophages, for example, only escape from the host after cell lysis, and these viruses have evolved special mechanisms to break down the cell wall and membrane when virus growth is completed. Cells infected with many animal viruses do not lyse at the end of the virus growth period, and a contributory factor to their death is that during infection these viruses selectively and irreversibly turn off host DNA, RNA and protein synthesis in favour of their own macromolecular synthesis. Many bacteriophages also turn off host macro-

molecular synthesis and some even cause the breakdown of host components to provide building blocks for their own synthesis. Productive infections of cells without cell death do occur, however. The viruses causing this type of infection have evolved methods of virion release which do not cause irreversible cell damage. Many animal viruses, for example, are released from the cell membrane by a budding process, during which the virus incorporates a small piece of modified cell membrane as one of its structural components. This type of infection is compatible not only with cell survival but also with simultaneous growth of the infected cell and virus production.

In chronological order the events which take place during the virus growth cycle are:

(1) Adsorption of the virion to the cell surface.

(2) Entry of the virus nucleic acid (bacteriophages) or whole or part of the virion (animal viruses) into the cell.

(3) Transcription and translation of viral mRNA from the virus genome.

(4) Genome replication.

(5) Assembly of progeny virions and their release from the cell.

The whole process is known as the virus growth cycle. The growth cycles of bacteriophages are usually short, being measured in minutes, whereas animal viruses have much longer growth cycles (Fig. 1.1). This is probably related to the faster metabolic processes found in bacteria, which, of course, have much shorter growth cycles than eukaryotic cells.

Not all virus infections lead to progeny production. Some of the reasons for this are self-evident, including host resistance (host-virus relationships are often highly specific, e.g. poliovirus will not infect non-primate cells because they lack surface receptors for virus adsorption). In other situations, however, the virus genome may enter the cell and remain

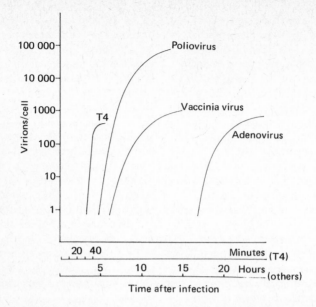

**Fig 1.1** Growth curves of animal and bacterial viruses.

there, sometimes for many cell generations, without concomitant virus production. The classical example of this type of host-virus interaction is the lysogeny of bacteria by bacteriophages. Here, after infection, the virus DNA inserts itself into and becomes covalently linked to the bacterial chromosome. An efficient mechanism under virus control then ensures that this situation is maintained until certain environmental conditions occur, when the virus genome is excised from the host chromosome and virion production and cell lysis occur.

Many animal viruses can also integrate their nucleic acid into the host cell genome. With many DNA viruses this may happen during the course of a productive infection. Rarely, and usually only under special conditions in the laboratory, integration of the DNA of these viruses may be accompanied by striking changes in cell morphology and growth patterns (transformation); such cells may cause tumours in susceptible hosts (Chapter 5). Many types of RNA tumour virus, however, transform cells efficiently and continue to grow in the trans-formed cells. The genome of these cells contain covalently linked DNA copies of the RNA virus genome.

# 2 The virion

## 2.1 Principles of virion construction (Table 2.1)

Two basic patterns of virus structure can be recognized: (a) protein subunits arranged in a spherical shell with cubic symmetry (a crystallographic term indicating the relationship of subunits to each other) and (b) protein subunits arranged with helical symmetry. In both cases the protein subunits surround the nucleic acid molecule to form a structure known as a nucleocapsid. Helical nucleocapsids of animal viruses are always enveloped in a membrane which contains virus proteins and host cell lipids; many plant viruses and some bacteriophages exist as naked helical nucleo-capsids. Only one group of viruses with cubic symmetry, the herpesviruses, is enveloped. In two groups of enveloped viruses (the toga-viruses and the RNA tumour viruses) the arrangement of the protein subunits of the nucleocapsid is not known. In many groups of viruses with cubic symmetry, the spherical shell is not in direct contact with the nucleic acid but encloses another protein structure containing the viral genome; this is known as the virus core.

Some very large viruses, including the T-even bacteriophages and the poxviruses, have been classified as 'complex', the assumption being that they have neither cubic nor helical symmetry. It now seems likely that the head of T-even bacteriophages is made up of protein subunits arranged with cubic symmetry. The arrangement of protein subunits in the core of poxviruses is unclear at present.

## 2.2 Special features of virion anatomy

### 2.2.1 Particles with cubic symmetry

All viruses with this basic structure that have been critically examined have icosahedral symmetry. (An icosahedron is a symmetrical polyhedron with 12 vertices and 20 faces, each an equilateral triangle which in turn may be further subdivided into small equilateral triangles.) Caspar and Klug have shown that this type of construction is the only general way in which spherical shells may be constructed from large numbers of identical protein subunits. Although it is impossible to put more than 60 identical subunits on the surface of a sphere in such a way that each is identically arranged. many viruses with icosahedral symmetry have more than 60 subunits in their shell. Caspar and Klug resolved this puzzle by invoking the principle of quasi-equivalence which allows the formation of these shells if the bonds holding the subunits together are deformed in slightly different ways in different parts of the shell. It is also possible to construct such icosahedra by using more than one kind of structural subunit. Subunits are often arranged in groups of 5 (pentons) and groups of 6 (hexons) in the shell; the adenovirus is a good example of a virus with this type of structure (Fig. 2.1). This is a well characterized virus in structural terms, largely because its structural subunits are soluble under mild conditions, a property not shown by the structural subunits of most other viruses. The adenovirus subunits can be readily purified from infected cells where they are made in large

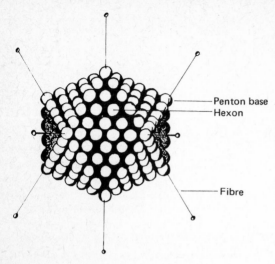

**Fig**. 2.1 Adenovirus – drawing of the virion.

Penton base
Hexon

Fibre

molecular weight of 60 000–65 000. Penton bases are difficult to work with but appear to be made of a number of identical polypeptides, molecular weight 70 000. The core contains two polypeptides (molecular weights 17 000 and 45 000) present in multiple copies.

The long penton fibres with their terminal knob are a feature unique to the adenovirus group. Distinguishing features of other icosahedral viruses include the number of structural subunits in the outer shell, which varies from 12 (bacteriophage $\phi$X174) to 812 (Tipula iridescent virus), their diameter, which varies from about 20 nm (parvoviruses) to 130 nm (Tipula iridescent virus), and the nature of the nucleic acid contained within the shell. This may be double stranded DNA (papova-, adeno-, herpes- and Tipula iridescent viruses), single-stranded DNA (bacteriophage $\phi$X174, parvoviruses), single-stranded RNA (picornaviruses, bacteriophage Q$\beta$ and its relatives, and many plant viruses) or double-stranded RNA (reoviruses). It is possible that many tailed bacteriophages have a head constructed with icosahedral symmetry; these viruses present so many other specialized features that they will be described in detail later.

Reoviruses stand out from the other icosahedral viruses in many ways. For example, they are the only group with double-stranded virion RNA, and their genomes are fragmented, each virion containing a number of different pieces of unequal sized RNA. In addition they are the only icosahedral viruses known which possess two shells, one surrounding the other. The inner shell contains the viral RNA and is known as the core It contains a transcriptase which synthesizes single-stranded RNA from the double-stranded template. Transcriptases from other viruses with icosahedral symmetry have not been described; this finding is probably connected with the observation that reoviruses are the only icosahedral viruses from which infectious nucleic acid has not been extracted.

amounts. The virus shell (capsid) is comprised of 252 structural subunits (capsomers), 240 being hexons with 6 neighbours, and a molecular weight of about 350 000 and 12 being pentons having 5 neighbours and a molecular weight 400 000–515 000. These occur at the vertices of the icosahedron, and consist of a base and a fibre with a terminal knob projecting from the surface of the virus. If the virus is disrupted by strong detergents or urea the hexons are released in two forms, those neighbouring pentons being solubilized as singles and those forming the triangular facets of the icosahedron being released as groups of 9 (nonamers). Contained within the virus is a core containing a duplex DNA molecule (molecular weight $20\text{-}25 \times 10^6$) associated with protein. Further complexities of structure are revealed when the polypeptide composition of the structural subunits is examined. Hexons appear to be made up of 3 identical polypeptides (molecular weight 120 000); fibres are also trimers, the subunit polypeptide having a

### 2.2.2 Filamentous viruses with helical symmetry

The classic example of a virus with this type of structure is Tobacco mosaic virus (TMV). The protein subunits of TMV interact to form a helical rod with a central hole. The RNA molecule (single stranded, molecular weight $2 \times 10^6$) runs between the subunits (not down the central hole) and its length determines the length of the particle. The molecular weight of the single subunit polypeptide is 17 000 and each particle contains 2 200 subunits. TMV is structurally one of the simplest viruses known.

### 2.2.3 Viruses with a lipid-containing membrane containing virus-induced proteins (envelope)

Three basic types of virus particle fall into this category: (i) Particles with an envelope which surrounds a helical nucleoprotein structure: (ii) Particles similar to (i) but with a nucleoprotein core not showing obvious icosahedral or helical symmetry: (iii) Particles with icosahedral symmetry, or with a complex structure, surrounded by an envelope. The former arrangement is typical of the herpesviruses, and the latter is shown by some poxviruses. Viruses in groups (i) and (ii) have RNA genomes; those in group (iii) have DNA genomes.

One of the most intensively studied viruses included in group (i) is vesicular stomatitis virus (VSV). This is the prototype virus of the rhabdovirus group, which contains vertebrate, invertebrate and plant viruses. VSV is a bullet-shaped virus (Fig. 2.2). Inserted into the outer surface envelope of the virus particle are a large number of glycoprotein (G) subunits. VSV also contains another protein (M) associated with the envelope; this is not exposed to the external environment of the virus, is not glycosylated and is thought to join the envelope to the helically arranged nucleoprotein (nucleocapsid) which is coiled within the envelope. The nucleocapsid uncoils readily

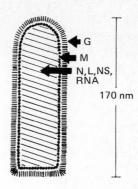

Fig. 2.2 Vesicular stomatitis virus — diagram illustrating the structural relationships between the different component polypeptides.

on virus disruption to reveal a thread-like structure. The detailed arrangement of the single-stranded RNA (molecular weight about $4.0 \times 10^6$), accounting for 1–3% of the total virion mass) and the one major (N) and two minor (L, NS) protein components of the nucleocapsid are not known. Isolated purified nucleocapsids are infectious, albeit with low efficiency compared to viruses, and contain transcriptase activity. The total molecular weights of the polypeptides contained within the virion account for practically all the coding capacity of the virus genome (total polypeptide molecular weight 383 000 made up of L 190 000, G 69 000, N 50 000, NS 40 000 – 45 000, M 29 000). The basic structure of the varions of myxoviruses and paramyxoviruses resembles that of VSV in many particulars, including a helical nucleocapsid with RNA polymerase activity, an envelope containing glycoproteins exposed at the surface of the virion and an envelope-associated protein similar to the M protein of VSV. Unlike VSV, however, these viruses are markedly pleomorphic in shape, being either roughly spherical or filamentous, and their envelopes contain additional surface proteins — one of them the enzyme neuraminidase. An important distinction between the myxoviruses and the paramyxoviruses and rhabdoviruses is found in the RNA, which in the myxoviruses is present in each virion in several fragments of differing size. The virions of the other groups each have a single unfragmented molecule.

Two groups of viruses fall into category (ii) – enveloped viruses with a nucleoprotein core of undetermined symmetry. These are the togaviruses and the RNA tumour viruses – totally unrelated viruses, which we shall consider separately. Togaviruses have a simple structure consisting of an envelope with one or two glycoprotein subunits present in multiple copies, and a single stranded RNA molecule, molecular weight $4 \times 10^6$, associated with another protein. Transcriptase and neuraminidase have not been detected in togavirus particles and virion RNA is infectious when extracted in a protein-free state.

RNA tumour viruses have a complex structure. Well defined is an outer envelope with glycoprotein subunits, and an inner core, or nucleoid, which contains a fragmented single-stranded RNA genome. The arrangement of the structures inbetween the envelope and the nucleoid is unclear as is the structure of the core itself. RNA tumour virus particles contain several enzyme activities, notably reverse transcriptase (see Chapter 5) and the RNA is not infectious when extracted from the virion.

With few exceptions, enveloped viruses mature at cell membranes, often by a budding process which culminates in the enclosure of the virus nucleoprotein helix or core by virus-modified cell membrane. The lipid composition of these viruses, not surprisingly, resembles closely that of cell membranes, and it seems that the capacity of the virus to control the lipid composition of its envelope is very limited.

### 2.2.4 Tailed bacteriophages

Bacteriophage T4 has been studied intensively for many years and its structure is understood

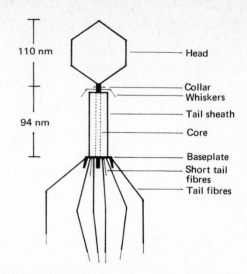

Fig. 2.3 Phage T4 – diagram of the virion

in some detail (Fig. 2.3). The head contains a linear, uninterrupted duplex DNA molecule. At least 11 polypeptides are found in purified heads (including collars and whiskers); the major structural polypeptide has a molecular weight of 45 000 and comprises 50% of the mass of the virion. How the DNA is packaged into the head is a fascinating and unsolved problem. Three different proteins and considerable amounts of the polyamines, spermidine and putrescine are found inside the head; none of these is absolutely necessary for infectivity, although one internal protein is needed for efficient head assembly and another, which is injected along with the DNA into the host, is needed for phage growth in some strains of *E. coli*.

The phage tail is made up of a hollow central tube surrounded by a contractile sheath which is joined to the head by a connector. At the other end of the tail is a hexagonal baseplate which has six tail pins. The whole structure is made up of at least 15 different polypeptides.

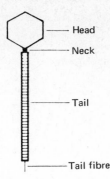

**Fig.** 2.4 Phage λ — diagram of the virion

The tail fibres are joined to the baseplate; the long fibres are concerned with phage adsorption and the short tail fibres probably hold the phage fast to the host during sheath contraction and DNA injection.

The structure of many tailed phages is less complex than that of phage T4. The virion of phage λ, for example (Fig. 2.4) consists of a DNA-filled head, probably constructed with icosahedral symmetry, which is joined by a narrow neck to a hollow tail built of subunits arranged as stacked discs. The tail terminates with a single tail fibre and there is no contractile sheath.

### 2.2.5 Poxviruses

These are large viruses with a correspondingly large genome — each virion contains a duplex DNA molecule with a molecular weight of approximately $160 \times 10^6$. Over 30 different polypeptides have been found in purified virions, and the detailed arrangement of these is poorly understood at present. Although poxviruses do not mature at cell membranes their virions contain small amounts of lipid and at least one glycoprotein. These components together with other polypeptides form a layer around the virus core which in thin sections of mature virions appears to be constricted centrally by two lateral bodies. The composition and function of these structures is unknown. The core contains transcriptase activity and other enzymes concerned with nucleic acid metabolism — under appropriate conditions isolated cores actively secrete newly-synthesized RNA into the surrounding medium. Recent evidence suggests that in some mammalian poxviruses a small proportion of mature particles may obtain an additional envelope at the cell surface; in general, however poxviruses remain largely cell associated at the end of the virus growth cycle. The mature virions of many types of poxviruses become embedded in a proteinaceous matrix (sometimes with a crystalline substructure, as in the case of insect poxviruses) to form large inclusion bodies which are readily seen by light microscopy. This type of inclusion body is also characteristic of many other insect viruses not belonging to the poxvirus group. In some groups (polyhedroses) many particles are contained in one inclusion; in others (granuloses) each virion is enclosed in a separate inclusion.

### 2.3 Viral nucleic acids

From the foregoing discussion two major facts concerning nucleic acids of virions are apparent. Firstly, the nucleic acid component is located internally within the capsid proteins. Secondly, that either DNA or RNA can act as the permanent repository for the genetic information thus bridging the informational gap between one vegetative replication cycle and the next. In almost all known plant viruses the nucleic acid is RNA, for animal and bacterial viruses it can be either RNA or DNA. Some further general statements can also be made; (1) viral RNA molecules are usually single-stranded and viral DNA molecules usually adopt the double-stranded configuration; (2) the larger viral nucleic acid molecules are usually DNA whereas the smaller viral genomes can be either DNA or RNA; and (3) for most viruses the nucleic acid component consists of a single molecule — in genetic terms most viruses therefore are haploid with a single chromosome. Within this general framework there is considerable variety and some exceptions to the general pattern.

The nucleic acid content of the virion ranges from about 1% to about 50%. The higher values are more characteristic of the large DNA phages such as the T-phages, and the low values, as in the case of animal viruses, usually indicate the presence of an envelope surrounding the nucleocapsid, e.g influenza, herpes simplex and RNA tumour viruses.

Viral RNA molecules range in molecular weight from about $10^6$ to $10^7$ daltons, whereas, for DNA viruses, the limits range from $10^6$ to about $16 \times 10^7$ daltons. While it is difficult to determine what genetic significance these values have in terms of the numbers of polypeptides coded, it is useful for comparative purposes to make simple calculations about the gene contents of genomes at the lower and upper ends of the molecular weight range. If we assume that the complete genome is poly-peptide-specifying (while not true this is probably close to being so) and that an average polypeptide contains 200 amino acid residues, then with the triplet code the average gene will contain 600 bases (600 base pairs if the nucleic acid is double-stranded). Therefore a small single stranded genome of $10^6$ daltons molecular weight such as that of the RNA phage MS-2, would contain sufficient information for 5 proteins (in fact the number is known to be 3). For a small double-stranded DNA genome such as polyoma, with a molecular weight of $3.0 \times 10^6$, this value is 8. At the upper end of the scale the number of proteins coded for by phage T2 ($2 \times 10^5$ base pairs) could be over 300, and for poxviruses it could rise to about 400. The reason for this exercise is to indicate that at one extreme the addition to a cell of genetic information to the tune of 3—5 genes is sufficient to cause the death of that cell and in the process synthesize several hundred new virus particles. It is equally clear that a virus such as phage T2 which has a hundred times the coding capacity, can only achieve the same end result. The question is then raised — what does T2 do with all this extra information?

While the majority of viruses contain a single nucleic acid molecule others contain more than one. For example, under some conditions the RNA of influenza virus can be obtained as 5 or more components of varying molecular weight; similarly the double-stranded RNA of reovirus has been recognized to exist in 3 class sizes with a total of 10 pieces plus some single stranded molecules. There is no base sequence homology between different genome segments and these can be regarded as being equivalent to individual virus genes. A different situation exists with a few plant viruses such as alfalfa mosaic virus where the particles are heterogeneous and the different components contain RNA molecules of different sizes. Apparently, infectivity requires a combination of the various components, single components being non-infectious. Fragmented or multicomponent genomes appear to be restricted to RNA viruses and neither characteristic has yet been identified for a bacterial virus. The genomes of all RNA viruses appear to be linear.

Considerably more information is available on the structure of viral DNA molecules and a variety of fascinating properties have been revealed over the last 10 years. The small single stranded DNA viruses, e.g. the icosahedral viruses $\phi$X 174 and S13 and the rod shaped viruses fl, fd and M13, have circular genomes. Circularity is also a property of the duplex DNA genomes of the polyoma and papilloma sub-groups of the papovaviruses. In fact, three forms of DNA can be isolated from polyoma virus. All have a molecular weight of $3 \times 10^6$ daltons but only one form, component III, is linear. The circular forms, components I and II, differ in several properties resulting from the fact that each of the two strands of form I are completely covalently bonded whereas at least one of the strands of component II is interrupted. As a consequence, component I may assume in solution a supercoiled configuration due to a shortage of turns in the helix — one single 'nick' would release this tension and convert the supercoiled form into an open 'relaxed' circle characteristic of form II. Covalently closed circular DNA molecules also appear as intracellular forms of some viruses having linear virion genomes, e.g. phages λ and P22 (see Chapter 3).

Circular virion DNA molecules are, however, less common than linear structures. Detailed study of these structures has revealed considerable anatomical variation. The DNA molecules extracted from a phage such as λ are all identical in respect to the sequence of bases along the molecule — if the sequence is read starting from the left end of the molecule this sequence is unique and would be repeated for all other λ molecules. This unique arrangement is found for other unrelated phages such as T1, T7, T5 and SP105. An alternative sequence arrangement was first found in phage T2 and is now known for other phage DNA's. The base sequences of a collection of T2 DNA molecules represent circular permutations of the common sequence — this can be envisaged by converting a series of circular molecules into linear molecules by breaking each once at a random location. It should be emphasized, however, that there is no evidence to suggest that the DNA of these phages is ever circular in the virion.

In all cases studied, a sequence of bases at the ends of linear molecules is repeated. In one form, exemplified by phage λ, the 5′ ends of the molecules are single-stranded for a length of 12 bases and the sequences are complementary. Under the appropriate annealing conditions the two ends of the molecule will hybridize to produce a circular structure. Since all λ DNA molecules are identical, the cohesive ends of one molecule will also hybridize to those of other molecules to form multimeric structures. Similarly, the ends of T7 DNA molecules will unite to produce circular molecules but only after partial digestion of the ends with a single-stranded specific exonuclease such as exonuclease III which hydrolyses the 3′ terminated chains. The need to produce single-stranded ends chemically in order to achieve circularization means that the terminal sequences are identical and duplex. This feature of terminal repetition of base sequences has proved to be common and the length of the redundant sequence varies from about 250 bases for T3 and T7 (less than 1% of the total sequence) to about 4000 for T2 (1—3% of the molecule). For phages T5 and T1 the redundant length forms about 8% and 10% of the molecular length respectively. The DNA molecules from the double-stranded adenovirus and the single-stranded adeno-associated virus reflect another type of variation in terminally redundant sequences — in both cases the single strands are capable of circularization and the evidence points to the repetition being inverted to produce a single-stranded circle with a duplex T-junction at the site of the hydrogen-bonded ends.

A further feature, observed initially with phage T5 and now repeated for phage SP50 and herpes simplex virus, is the presence of a small number of single strand interruptions along the length of the molecule. For T5, which is unique, the nicks are specifically located and are restricted to one strand whereas for SP50, another unique molecule, they appear to be randomly situated.

Clearly these various physical pecularities must have some biological significance and this is discussed in later chapters dealing with the replication and recombination of DNA molecules.

## 2.4 The virion: function
The functions of the virion are (1) to permit the transmission of infection through a hostile environment; (2) to maximize the probability of an efficient iniation of infection and, in some cases, to play a role in subsequent events of the virus growth cycle. Our description of virion structure has shown that viruses have evolved several different types of particle to serve these ends.

### 2.4.1 Transmission of infection
Viruses are not motile and hence rely entirely on passive means for the transmission of infection from cell to cell. During their extra-cellular existence, virus particles are exposed to environmental factors which may be hostile to their survival as infectious agents. These include heat, drying, radiation, extremes of pH, enzymes (e.g. nucleases, proteases), and specific antibodies. Despite their simple structure and limited genome size many viruses have shown considerable versatility in overcoming these hazards.

### 2.4.2 Initiation of infection
The role virions play here is (1) to adsorb to receptor sites on the surface of the cell; (2) to facilitate entry of the virus genome into the cell (in the case of many animal viruses the entire virion enters the cell); (3) in some cases to play a part in subsequent events of the virus growth cycle.

### 2.4.3 Adsorption
Examples of specialized virion structures concerned with virus adsorption include the tail fibres of T-even bacteriophages, the pentons of adenoviruses, and the haemagglutinin spikes on the surface of influenza virus. These structures all attach to specific receptors on cell surfaces. Many animal and plant viruses do not show such morphologically distinct structures but are still capable of interacting with specific cell receptor sites — a good example is poliovirus, whose growth is restricted to primate cells solely on the basis of adsorption specificity, illustrated by the wide spectrum of non-primate cell types susceptible to infection with naked poliovirus RNA.

The tail fibres of T-even bacteriophages have been extensively studied (Fig. 2.3) and the process of adsorption by these viruses is well characterized. The time course of phage adsorption follows first-order kinetics with respect to the concentrations of both bacteria and phage; the reaction is sometimes reversible at low cation concentrations, but at higher ionic concentrations this reversible adsorption is obscured by a fast, irreversible, fixation. Magnesium ions and tryptophan are required for extension of phage T4 tail fibres and for adsorption: in the absence of these co-factors tail fibres are folded against the tail and adsorption does not occur. Each bacterium has many sites at which phage can adsorb. It is easy to isolate phage-resistant bacterial strains (see Chapter 4); most of these turn out to have altered receptor sites, to which the phage can no longer adsorb. Phage mutants able to grow on these strains can be isolated and the mutations can be mapped within tail fibre gene 37. The molecular basis of the interaction between the tail fibre and the bacterial receptor is not understood, but an analogy may be drawn with antibody/antigen and enzyme/substrate interactions.

Many of the conditions for optimal adsorption of animal viruses are similar to those described for phage; in general the attachment rate is relatively insensitive to temperature and is maximal in the presence of divalent cations. Co-factor requirements have not been demonstrated. Many animal viruses agglutinate red cells and this phenomenon has been used as a model system in the study of virus adsorption.

Haemagglutination by viruses is also of great practical importance, as it can be used to detect viruses, to titrate them and to assay anti-viral antibodies. Virion structures which adsorb to red cells and cause haemagglutination include the haemagglutinin spikes projecting from the surface of influenza virus (one of the virion glycoproteins) and the fibres of adenoviruses.

## 2.4.4 Entry

Bacterial and animal viruses have adopted completely different mechanisms for the entry of their genomes in the host cell, probably due to the completely different properties of the thick, rigid bacterial cell wall and the thin, mobile animal cell membrane. Bacteriophage virions often contain complicated structures concerned with the injection of nucleic acid directly into the cytoplasm of host bacteria — no such structures have been described in animal virus particles, which are often engulfed entire into the host cell and subsequently uncoated. The T-even phages, for example, attach to the host, first by the tail fibres and then by the basal plate. A complex series of events then follows, including localized lysis of the cell wall (probably by a phage enzyme) and the contraction of the tail sheath which forces the hollow inner tail core through the cell wall. Subsequently the phage DNA is injected into the cytoplasm of the bacterium.

The ability to cause cell fusion is a property shown by the virions of some animal viruses, and it is possible that this phenomenon is connected with entry (Chapter 3). Studies on virus-induced cell fusion have usually employed Sendai virus, a paramyxovirus. These experiments have shown that when large amounts of this virus are applied to susceptible cells, their cytoplasmic membranes rapidly fuse, with the production of multinucleate giant cells. Fusion is mediated by components of the virus envelope and is temperature dependent. Virus infectivity is not required.

## 2.4.5 Functions of the virion after entry — virion enzymes

The first enzyme to be discovered in virus particles was the neuraminidase of influenza virus. This enzyme is also found in the virions of paramyxoviruses. Its action is to cleave the glycosidic bond joining the keto group of neuraminic acid to D-galactose or D-galactosamine; it modifies the carbohydrate moiety of the virion haemagglutin by catalysing the removal of terminal sialic acid residues.

For many years neuraminidase remained the sole example of a virion enzyme, and animal virologists held to the general rule that virions were concerned solely with the protection of the virus genome and with its entry into the cell — ideas largely derived from work on bacteriophages. In 1967, however, Kates and McAuslan found that the virions of vaccinia virus contained DNA-dependent RNA polymerase activity, and since then numerous other animal, plant and insect viruses have been shown to possess DNA or RNA polymerases in their virions (Table 2.1). In general these enzymes, known collectively as transcriptases, synthesize single-stranded messenger RNA from the virion nucleic acid, which may be double-stranded DNA or single- or double-stranded RNA. The reverse transcriptase of RNA tumour viruses, however, synthesizes DNA from a single-stranded RNA template (Chapter 5). The role these enzymes play during infection will be considered in Chapter 3; that they are essential for virus growth is shown by the observation that protein-free intact nucleic acid molecules extracted from virions possessing transcriptases are *not* infectious, contrasting with the ready isolation of infectious nucleic acid from the virions of viruses which lack such enzymes e.g. picornaviruses, togaviruses, papovaviruses and adenoviruses. Virion transcriptases have in general proved to be rather difficult enzymes to work with, and the reverse transcriptase of RNA tumour viruses is the only example so far

**Table 2.1** Summary of virion properties of some major RNA virus groups

| Virus group | Genome | Nucleocapsid symmetry | Envelope | Transcriptase | Number of different polypeptides |
|---|---|---|---|---|---|
| Spherical bacteriophages (Qβ) | S,[a] +,[b] 1[c] | Cubic | — | — | 2 |
| Spherical plant viruses | S, +, 1 | Cubic | — | — | 1 |
| Filamentous plant viruses (TMV) | S, +, 2 | Helical | — | — | 1 |
| Picornaviruses (poliovirus) | S, +, 2.6–2.8 | Cubic | — | — | 4 |
| Togaviruses | S, +, 4 | Unknown | + | — | 3 |
| RNA tumour viruses | S, +, (10) | Unknown | + | + (reverse) | 7 |
| Rhabdoviruses | S, −, 4 | Helical | + | + | 5 |
| Paramyxoviruses | S, −, 7 | Helical | + | + | 6 |
| Myxoviruses | S, −(3−5) | Helical | + | + | 6 |
| Reoviruses | D, (15) | Cubic | — | + | 7 |

[a] S, single-stranded; D, double-stranded
[b] Polarity of genome strand; + strands are mRNA-like
[c] Molecular weight $\times 10^{6}$ daltons; viruses with fragmented genomes have this figure in parenthesis.

that has been removed from virions and purified in an active state. The virion transcriptases of the remaining viruses are found in helical nucleocapsids (rhabdoviruses, paramyxoviruses and myxoviruses) or virion cores (reoviruses and poxviruses) and it is entirely possible that the structural integrity of these virion substructures is required for correct enzyme function.

A considerable number of other enzyme activities have also been found associated with the virions of animal viruses. It is often difficult to decide whether these enzymes are of host origin and simply remain associated with virus particles during purification, or whether they are an integral part of the virion structure. Their function during the virus growth cycle is at present unclear.

**Table 2.1** Summary of virion properties of some major DNA virus groups

| Virus group | Genome | Nucleocapsid symmetry | Transcriptase | Number of different polypeptides |
|---|---|---|---|---|
| Filamentous bacteriophages (fd) | S,[a] C,[b] 2[c] | Helical | − | 2 |
| Spherical bacteriophages ($\phi$X174) | S, C, 1.7 | Cubic 12 capsomers | − | 5 |
| Bacteriophages with non-contractile tails ($\lambda$) | D, L, 32 | Complex | − | 12 |
| Bacteriophages with contractile tails (T2) | D, L, 130 | Complex | − | 30 |
| Papovaviruses | D, C, 3−5 | Cubic, 72 capsomers | − | 6 |
| Adenoviruses | D, L, 20−25 | Cubic, 252 capsomers | − | 13 |
| Herpesviruses | D, L, 100 | Cubic, 162 capsomers | − | 30 |
| Poxviruses | D, L, 160 | Complex | + | >30 |

[a]   S, single-stranded; D, double-stranded
[b]   C, circular; L, linear
[c]   molecular weight x $10^6$ daltons

# 3 The virus-infected cell

In general the events which take place in the virus-infected cell can be divided into four basic stages: transcription of the virus genome, translation of the viral mRNA, genome synthesis, and assembly of progeny virus particles. Some viruses have combined the first and third of these stages as their single-stranded RNA genomes act as mRNA (picornaviruses, togaviruses, RNA phages), and a few plant viruses have virtually dispensed with the last stage as infections under natural conditions appear to be transmitted by naked nucleic acid molecules.

Bacteriophage genomes enter host cells as naked nucleic acid molecules; this is not the case with animal viruses although partial or complete uncoating of virus nucleic acid often occurs before the start of virus-directed macromolecular synthesis in the infected cell.

It seems probable that most animal virus particles can enter the cell by a process known as viropexis – the phagocytosis of virus particles into cytoplasmic vacuoles. Subsequent breakdown of the vacuole membrane follows with release of virus particles into the cytoplasm. Some viruses with icosahedral shells have been shown to enter cells by direct penetration of the plasma membrane. There is also clear evidence that some enveloped viruses achieve both entry and uncoating by fusion of the outer membranous coat of the virion with the cell plasma membrane. The nucleoprotein core or helix of the virus is thus discharged into the cytoplasm where virion transcriptases or mRNA-like genomes find an environment suitable for subsequent biosynthetic events.

The initial stages of virus uncoating are almost certainly brought about by pre-existing cell enzymes. Partially uncoated particles (cores) of icosahedral viruses with DNA genomes traverse the cytoplasm and enter the nucleus; the nature of the transport mechanism involved is obscure.

## 3.1 Transcription and translation in the virus-infected cell

All viruses must synthesize mRNA. The central role of mRNA derives from the fact that viruses use cellular ribosomes and soluble factors to translate their mRNA. Viruses can be grouped according to their pathways of mRNA synthesis (Fig. 3.1) and we have used this classification, proposed by Baltimore, in the account which follows.

*3.1.1 Double-stranded DNA viruses (Group I)*
This is a heterogeneous group, containing the T-even and T-odd and λ bacteriophages of *E. coli* and the animal viruses belonging to papova- adeno-, herpes- and poxvirus groups. Only the poxviruses have a virion-associated transcriptase; infectious DNA can be extracted from the virions of all the other quoted examples, and these viruses use host transcriptases during at least the early stages of their growth cycles. A characteristic of many of the viruses in this group is the division of transcriptional events into at least two phases –

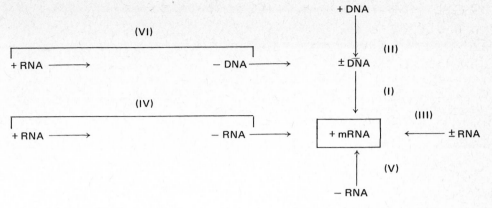

I Double-stranded DNA viruses
II Single-stranded DNA viruses
III Double-stranded RNA viruses
IV Single-stranded RNA viruses, mRNA
identical in base sequence to virion RNA

V Single-stranded RNA genome complementary
in sequence to mRNA
VI Single-stranded RNA genome with a DNA
intermediate in their growth.

Fig. 3.1 Composite diagram of pathways of mRNA synthesis used by various classes of viruses. mRNA defined as +RNA. (after Baltimore).

early and late. Early transcription occurs before the onset of virus DNA synthesis, and many of the mRNA species produced code for enzymes concerned with virus nucleic acid synthesis. Late transcription is dependent on prior, and in some cases ongoing, virus DNA synthesis and the mRNAs produced code largely for virion structural proteins. In many cases, transcription takes place from different strands of the virus DNA at different times in the virus growth cycle.

Special features of transcription and translation are now illustrated by reference to individual groups of viruses.

*(a) Bacteriophage T7.* Phage T7 illustrates a relatively straightforward pattern of replication. The genome of this virulent phage codes for about 30 proteins. Mutations, identifying some 25 genes, map as a linear sequence which is co-linear with the non-permuted genome (Fig. 3.2). The gene order shows a marked

grouping of related functions, with the left hand 20% of the molecule specifying proteins mainly concerned with viral DNA synthesis, the remaining genes through to the right hand terminus being required for the synthesis of virion proteins and maturation.

Infection leads to the rapid inhibition of host mRNA synthesis and the appearance of T7-specific mRNA which continues to be made throughout infection. Host DNA is degraded to form a major precursor of viral DNA which, together with phage structural proteins, is synthesized from about 10 minutes after infection until lysis at 30 minutes. Virus assembly is initiated soon after the formation of the precursors.

Three classes of T7-specified proteins can be resolved according to their time of synthesis. Class I proteins are made from 4 to 8 minutes, Class II from 6 to 15 minutes and Class III from 8 minutes until lysis. The Class I proteins are

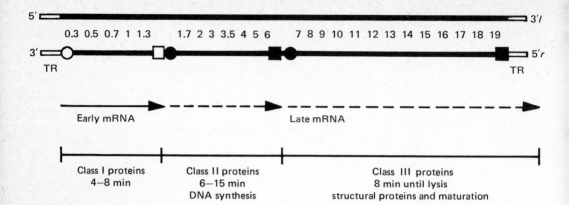

**Fig. 3.2** Chromosome of phage T7 showing arrangement and functions of genes and pattern of transcription. Open circle and square denote respectively the promotor and terminator locations for mRNA synthesized with the host cell RNA polymerase. Closed circles and squares represent promotors and terminators for T7 RNA polymerase (the number and the location of these sites is uncertain). TR is the terminal redundancy which is not transcribed.

coded for by a few genes, including gene 1, located at the left hand end of the genome. The DNA synthesis functions specify the Class II proteins and the Class III proteins, extending to the right hand end of the genome code for virion proteins and assembly.

Within 5 minutes after infection some 12 different species of T7 mRNA can normally be detected and represent the entire complement of T7 transcripts. All species of T7 mRNA are transcribed from the same DNA strand, referred to as the *r* strand. mRNA for Class I proteins correspond to 5 species and are defined as 'early' RNA since they are made in the absence of phage-coded protein synthesis. Conversely, the remaining RNA species, which require the synthesis of Class I phage-coded proteins, are termed 'late' RNA. The gene 1 product is the sole activity necessary to initiate late gene transcription. Gene 1 codes for an RNA polymerase, which is quite different from the bacterial RNA polymerase. With purified T7 DNA template the host enzyme synthesizes

predominantly early RNA whereas the T7 enzyme catalyses only late RNA synthesis.

Thus, upon infection the bacterial RNA polymerase binds to specific sites (promoters) present only in the early region of the T7 genome and initiates synthesis of early mRNA. Transcription is specifically terminated before entering the late region. Translation of early transcripts leads to synthesis of the new T7 RNA polymerase which recognizes only late promoters, thus initiating late transcription and subsequently synthesis of Class II and Class III proteins.

*(b) Bacteriophage T4.* T4, like T7, is a virulent phage with a basically similar pattern of replication. However, the T4 genome is much larger, containing information sufficient to specify about 150 proteins. A third of the coding capacity is required for structural protein synthesis and virion assembly and about 20 genes are essential for phage DNA synthesis. Some of the remaining information codes for functions which duplicate host functions,

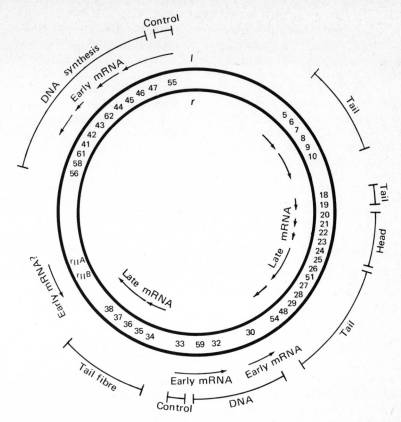

**Fig. 3.3** Partial genetic map of phage T4 showing functional grouping of genes and pattern of transcription of some map segments (arrows). Numbers represent individual genes.

particularly for nucleotide metabolism and transfer RNA synthesis, which are dispensable under laboratory conditions. The circular genetic map shows clustering of functionally related genes, although not to the marked degree seen for T7 (Fig. 3.3).

Virus-coded proteins can be divided into four main classes on a temporal basis. Classes A, B and C are made at different periods during the early stages of infection and include proteins which arrest host macromolecular synthesis and those required for DNA replication, while Class D proteins, which include

virion proteins, are made from mid-latent period until lysis.

Specific T4 mRNA molecules are also synthesized in a time-controlled manner, two or more classes being synthesized at early times. At later times, late mRNA species are detected. Early mRNA is transcribed predominantly from the *l* strand of the genome whereas most late transcription is from the *r* strand. There is good evidence that some mRNA molecules are polycistronic, i.e. are transcripts from contiguous blocks of genes (presumably initiated from a single promoter region). All this evidence strongly suggests that transcriptional control is a major feature of T4 regulation.

All T4 transcription is carried out by the host RNA polymerase and no evidence for the induction of a wholly new polymerase has emerged. However, it is likely that phage-specified proteins are also involved in the synthesis of T4 mRNA. Gene 33 and 55 products, for example, correspond to positive regulators of late gene transcription by permitting transcription of DNA sequences unavailable to the host RNA polymerase. The requirement for DNA synthesis may reflect differences between replicating and non-replicating molecules which preclude the latter from serving as template for late mRNA synthesis.

T4 infection also modifies the translational machinery of the cell, e.g. the phage codes for several species of transfer RNA. Therefore the possibility of translational control cannot be excluded but is probably of minor significance compared with transcriptional regulation.

(c) *Bacteriophage* λ. λ is a temperate phage, infection leading either to virus replication resulting in cell lysis (the only form of inter-action available to virulent phage) or to lysogeny, a non-lethal infection in which phage genes are not expressed and the phage genome persists and is passed to all daughter cells. The replication of temperate phages is subject to complex controls. In addition to the regulation of lytic development, controls are required to separate the mutually exclusive lytic and lysogenic responses. For the moment only the lytic cycle will be considered.

Virus replication requires the function of about 30 of the 40 known genes. The λ genetic map again reveals the highly ordered arrange-ment of genes with related functions, which is a characteristic of the larger phages (Fig. 3.4). The regulated sequence of events which occur during lytic infection are similar to those described for T4 and T7, although, as a rule, temperate phages do not cause the early inhibition of host macromolecular synthesis.

Hybridization studies with λ mRNA indicate that both DNA strands are transcribed. Early in infection mRNA is synthesized on both *l* and *r* strands from a region in the right half of the genome. At later times this early transcription is largely replaced by mRNA synthesis from the left half and extreme right portion of the *r* strand

Three classes of mRNA can be defined according to time of synthesis and dependence on the function of phage-induced regulatory proteins. There is no evidence for the synthesis of a phage-coded RNA polymerase and all λ transcription is catalysed by the bacterial enzyme. Early mRNA is transcribed leftwards on the *l* strand and rightwards on the *r* strand from two promoters, $P_L$ and $P_R$, located to the left and right of the CI gene. Early transcription from $P_L$ stops to the left of gene N and rightward transcription ends just beyond the *cro* gene. Delayed-early mRNA is also synthe-sized from $P_L$ and $P_R$ promoters but the transcripts include the region from the CIII to *int* genes on the left of CI and the CII to Q region to the right (this sequence includes mRNA for O and P gene products which are essential for λ DNA synthesis). The N gene product is specifically required for the synthesis of delayed-early mRNA and operates by preventing termination of early transcription.

Expression of late λ functions (structural protein synthesis and lysis) is dependent on the function of gene Q. λ DNA synthesis is not required and the N gene product, while necessary for the synthesis of gene Q mRNA, is not directly involved in late gene transcription. There is evidence suggesting that the Q gene product modifies the host RNA polymerase to stimulate late gene transcription. All late transcription is initiated at a single promoter site situated between the Q and S genes and the entire late region, genes S and R and A

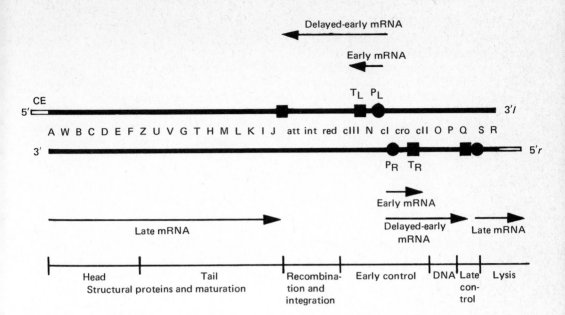

**Fig. 3.4** Chromosomes of phage λ showing arrangement and functions of genes and pattern of transcription (arrows). Circles and squares indicate promotors and terminators respectively. CE denotes cohesive end.

to J, is probably transcribed as a single poly-cistronic mRNA molecule which is subsequently cleaved, prior to translation. In any case the presence of only one late gene promoter must require fusion of the two ends of the molecule and in fact circularization of infecting genomes occurs soon after infection.

As with T4, the synthesis of early and delayed-early mRNA is switched off following the appearance of late gene transcripts. For λ this event is controlled by the *cro* gene product, which acts as a regulator by causing the inhibition of transcription.

*(d) Papova-, adeno- and herpesviruses.* These animal viruses are grouped together because their transcription occurs in the cell nucleus. Little is known about the enzymes involved; host transcriptases are probably used. The primary transcription product is processed by cleavage into smaller molecules and by the addition of a segment of polyadenylic acid. Shortly after adenylation the mRNAs leave the nucleus and associate with host ribosomes in the cytoplasm to form polyribosomes. Papova- and adenoviruses show clear patterns of early and late transcription; in SV40, about 30% of the genome is transcribed early, and 100% transcribed late (i.e. after DNA synthesis). In the case of adenoviruses, 10–20% of the genome is transcribed early and 80–100% late. Nuclear processing also plays an important role in the control of viral mRNA production as not all the products of transcription appear in the cytoplasm. A clear cut distinction between early and late transcriptional events cannot be made in the case of herpesviruses. This is also reflected in the pattern of protein synthesis shown by this group of viruses, many proteins

27

being made throughout infection without a clear early-late switch. Papovavirus and adenovirus structural proteins, on the other hand, are late proteins and are not made before the onset of virus DNA synthesis. The function of the early proteins made by these viruses, which includes T antigen, is unknown.

*(e) Poxviruses.* Poxviruses are unique amongst DNA viruses in their possession of a virion transcriptase. Their growth cycle occurs entirely in the cytoplasm of the host cell. After the first stage of virus uncoating, which leads to the production of cores, the virion transcriptase becomes active in the production of early mRNA. This mRNA is released from the virus core by an energy-requiring process; prior to this it is adenylated, possibly by the polyadenylate polymerase which is found in the virion core. Virus protein synthesis can be detected almost immediately after infection. Transcription of late mRNA is turned on shortly after the onset of virus DNA synthesis and does not occur if virus DNA synthesis is blocked by specific inhibitors; the transcriptase involved has not been identified but is almost certainly virus coded. Proteins coded for by early mRNA include enzymes such as thymidine kinase, DNA polymerase and DNAses and also some proteins found in the mature virion; however, most of the virus structural proteins are late. At about the time of onset of the synthesis of virus late proteins, the synthesis of most, but not all, early proteins is turned off; if virus DNA synthesis is blocked, however, early proteins continue to be made for many hours.

The synthesis of some late proteins is also turned off after a short time, although most are made for many hours. Transcription occurs in discrete areas of the cytoplasm, termed factories, but the mRNA diffuses rapidly throughout the whole cytoplasm of the cell, where protein synthesis takes place. The newly synthesized proteins rapidly move back to the factories where viral DNA synthesis and assembly occurs.

*3.1.2 Single-stranded DNA viruses (Group II)*
This group is exemplified by two classes of minute phage with circular genomes. The $\phi$X174 and S13 class have icosahedral virions whereas the second class, which includes phages M13, f1 and fd, are rod-shaped. With a coding capacity for only about 10 proteins these phages show a marked dependence on host functions; even so, host DNA synthesis is abruptly inhibited about half-way through the $\phi$X174 replication cycle. By contrast, bacteria infected with the filamentous phages continue to function and even to multiply. Filamentous phages are released by continual extrusion of virions through the cell wall rather than by lysis — an unusual phenomenon with phages.

Early after $\phi$X174 infection the single-stranded parental genomes are converted to duplex circles by synthesis of a complementary strand which serves as the template for all phage transcription. There is no evidence for division of mRNA synthesis into early and late species, all transcripts being detected at all times during replication. However, there is clear evidence for the sequential appearance of phage-coded proteins; the manufacture of virion proteins does not begin until mid-way through infection. At about the same time the pattern of DNA synthesis changes.

Analysis of $\phi$X174 mRNA synthesized *in vivo* has identified up to 20 separate species. Many are polycistronic, including molecules longer than the entire genome. The number and sizes of these species indicates the production of overlapping transcripts, i.e. a given genome sequence is represented by more than one species of mRNA molecule. It is believed that all $\phi$X174 mRNA is transcribed with an unmodified host RNA polymerase and that to account for the overlap of transcriptional sequences it would appear that some of the

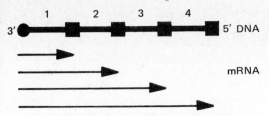

**Fig. 3.5** Scheme for overlapping transcription. The four DNA segments are transcribed from a single promotor (circle) with weak terminators (squares) at the end of each segment. mRNA for segment 1 is consequently more abundant than mRNA for segment 4.

termination sites are not completely effective and that a proportion of mRNA molecules continue synthesis through one termination site to the next terminator (Fig. 3.5). This particular mode of transcription would produce different amounts of mRNA for different genes. Translation of all mRNA species would result in an excess of some proteins relative to others, a situation which is characteristic of infection by $\phi$X174 and many other phages.

*3. 1. 3 Double-stranded RNA viruses (Group III)*
Reoviruses are the best understood viruses in this group, which also includes some insect and plant viruses. They all have a segmented genome, a virion-associated transcriptase, and multiply in the cytoplasm of the host cell. The virion transcriptase can be activated *in vitro* by treating virions with chymotrypsin. This removes the outer icosahedral shell leaving the inner core. The transcriptase remains firmly attached to the template and transcription occurs from the 10 genome segments by a conservative mechanism. Only one strand is transcribed. The 10 mRNAs produced are equivalent in size to the different genome segments and a direct relationship also exists between genome segment size and the molecular weight of newly synthesized poly-

peptides. The frequency of transcription of the different genome segments is controlled; the small segments are transcribed more often than the medium sized, which are in turn transcribed more often than the large segments. mRNA synthesis is first detected 2 to 4 hours after infection; an increase in rate of synthesis occurs after 4½ h, probably due to the synthesis of new transcriptase molecules.

Virus protein synthesis is first detected at about 3 hours after infection. The relative rate of formation of the virus-induced polypeptides remains similar throughout infection, and there are no distinct early and late polypeptides.

*3.1.4 Single-stranded RNA viruses whose mRNA is identical in base sequence to virion RNA. (Group IV)*
*(a) Bacteriophages.* All known RNA phages belong to this group, the best-studied examples being Qβ and a second class of related phages, R17, f2 and MS-2. These phages have much in common. Their linear genomes contain about 4000 bases and code for 3 proteins – a major coat protein, a minor coat protein (A protein) and a replicase gene which provides a subunit of the multicomponent enzyme required for replication of the phage genome. In addition, Qβ produces a fourth protein designated A1. The order of genes from the 5′ end of the RNA molecule is; A-coat-(A1)-replicase. Although single-stranded, a major portion of the genome is hydrogen-bonded into a series of hairpin-like loops. Short sequences of 50–150 bases occurring at either end of the molecule and between each of the 3 genes are not translated and are concerned with ribosome attachment and control of the translation process.

Viral replicase is synthesized early in infection and is switched off at later times, whereas coat protein and A protein are produced throughout infection. At all times the coat protein is made in a ten-fold excess over other phage proteins.

The phage genome is also the mRNA, and in an *in vitro* protein synthesizing system all phage genes are translated from the virion RNA molecule In this respect genome replication and transcription are synonymous. This process occurs first by synthesis of a complete RNA strand complementary in sequence to the genome; the complementary strand then serves as template for viral genome synthesis (see below). Thus there is no possibility of transcriptional control by the selective synthesis of particular regions of the genome. In fact all evidence indicates that gene expression is regulated by selective translation of the genome resulting from selective binding of ribosomes to the different protein synthesis initiation sites. This selection is influenced by the folded structure of the viral RNA which reduces the availability of certain ribosome binding sites. Thus ribosomes preferentially bind to the coat protein gene. Initiation of replicase synthesis requires translation of part of the coat protein region and it has been suggested that translation of the coat gene destroys hydrogen bonding at the initiation site of the replicase gene, which is consequently exposed for ribosome binding. The low rate of A gene translation can also be explained in a similar manner.

A phenomenon known as translational repression shuts-off replicase synthesis during the later stages of infection. Coat protein binds to a region of the genome which covers the end of the coat gene, the beginning of the replicase gene and the non-translated section between. As infection progresses, the coat protein subunits accumulate and the probability of a subunit binding to the genome, and so preventing translation of the replicase gene, increases.

*(b) Animal viruses – picornaviruses and togaviruses.* Poliovirus is the prototype member of the picornavirus group and is the best understood virus in this group. The genome of this virus acts as a single mRNA molecule, which is the template for the synthesis of a number of viral polypeptides. Thus, as with RNA phages, genome transcription and replication cannot be differentiated on the basis of complementarity between the viral mRNA and the genome. The RNA is infectious, is adenylated and contains the sequences necessary for ribosome binding and polypeptide initiation and termination. In contrast to the multiple initiation and termination sites on the polycistronic genomes of RNA phages, the poliovirus genome contains only one polypeptide initiation site and is translated as a monocistronic message. The resulting large protein, termed NCVPOO, has a molecular weight $> 200\ 000$ daltons and is cleaved during and after synthesis by proteases, possibly of host origin, to yield separate functional units. In this way the virus bypasses the apparent inability of mammalian cells to terminate and re-initiate translation at internal sites of mRNA molecules. The cleavage proceeds in at least three steps (Fig. 3.6). First, the single large polypeptide is cleaved during its synthesis into smaller polypeptides. This explains why the large polypeptide is never seen in infected cells; it never really exists under normal conditions because it is cleaved before its synthesis is completed. The second class of cleavage involves further processing of the initial products and occurs more slowly. These cleavages, for example, lead to the production of three capsid proteins VP-0, VP-1, and VP-3; the precursor of these proteins (NCVP 1) has a half-life of about 15 minutes. The final cleavage stage occurs late in virion maturation when one of the capsid proteins is cleaved into two pieces; this event happens at about the time of RNA packaging in the virion and can be compared to the assembly related cleavage of structural proteins which is a common feature of assembly of many other viruses.

Togavirus genomes are infectious when extracted from virions, are adenylated, and can presumably act as mRNA in the cell. The virion

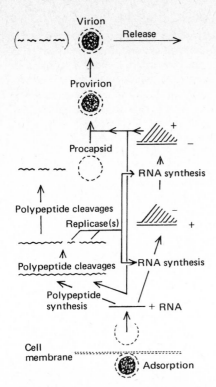

**Fig. 3.6** Diagram showing the sequence of events during the growth of poliovirus.

structural proteins are almost certainly derived from a high molecular weight precursor polypeptide by proteolytic cleavage, and the general pattern of virus macromolecular synthesis is assumed to resemble that of picornaviruses, although no giant precursor polypeptide equivalent to the poliovirus NCVP00 protein has yet been demonstrated.

### 3.1.5 Single stranded RNA viruses, genome complementary in sequence to mRNA (Group V)

Three groups of viruses fall into this category, the rhabdo- and paramyxoviruses with a single intact RNA genome, and the myxoviruses which have segmented genomes.

A transcriptase is associated with the helical nucleocapsid of the rhabdoviruses and paramyxoviruses. This enzyme synthesizes monocistronic mRNAs from the RNA template immediately after virus uncoating, the size of the mRNA molecules corresponding roughly to the size of the newly synthesized proteins. The mRNAs are adenylated. As far as is known the relative proportions of the different mRNAs synthesized does not change during infection, although the rate of synthesis of all mRNAs increases during infection due to the synthesis of new transcriptase molecules and templates. Likewise, the relative proportions of the various proteins synthesized does not change during infection. The growth of these viruses takes place entirely in the cytoplasm of the host cell and does not require the presence of a nucleus.

The influenza viruses are the sole members of the myxovirus group. Like the rhabdoviruses and paramyxoviruses, their nucleocapsids contain transcriptase activity, but they differ from these viruses in that early events essential for their growth take place in the cell nucleus. These events are blocked by actinomycin D, suggesting that host transcription may be involved. At present the site of virus transcription is not known. It is likely that transcription occurs from individual ribonucleoprotein complexes, each containing a distinct genome segment. A reasonable correlation can be made between the size of the various genome segments and the size of the newly synthesized virus polypeptides; approximately 10% of the total virus genome codes for non-structural proteins. Virus proteins are synthesized in the cytoplasm; some remain there, whereas others, notably the nucleoprotein, migrate into the nucleus where nucleocapsid assembly takes place. These structures then migrate out into the cytoplasm to take part in virus assembly at the outer cell membrane. The synthesis of a few non-structural proteins can

31

be detected before the onset of the major structural proteins; the function of these early proteins is unknown.

### 3.1.6 Single stranded RNA genome with a DNA intermediate in their growth. (Group VI)

RNA tumour viruses are the sole members of this group. See Chapter 5.

## 3.2 Post-translational modification of proteins

Many virus proteins are modified after their synthesis. An extreme example is the complicated processing, by proteolytic cleavage, of the picornavirus primary translation product, NCVPOO. Proteolytic cleavage of certain virus structural proteins during maturation is found in many virus groups, including phages T4 and λ (head proteins), adenoviruses, poxviruses (major core proteins), reoviruses, RNA tumour viruses and influenza viruses (haemagglutinin). In most of these cases the enzymes responsible have not been identified, although genetic data suggest virus coding in the case of phage T4. A host enzyme (serum plasmin) has been shown to be responsible for the cleavage of the precursor of the two haemagglutinin subunits of influenza virus. Here both cleavage products are incorporated into the virion; with many other viruses only one fragment of the cleaved protein is incorporated into the virion, the other probably being completely degraded by host proteases. Why proteins should be modified by cleavage during virion assembly is not understood.

Proteins exposed on the surface of enveloped viruses are usually glycosylated after their synthesis. The enzymes responsible are almost certainly host coded; the carbohydrate moeity of these proteins is usually small in size compared to the size of the protein (11% of the VSV glycoprotein is carbohydrate by weight) and is added soon after the completion of the synthesis of the protein. Its function is not clear but is almost certainly connected in some way with the location of the protein at a particular site in membranes. Even less well understood is the role of phosphorylated virus proteins; examples in this category include the fibre protein of adenoviruses, the minor ribonucleoprotein component NS of VSV, and a core component of vaccinia virus.

## 3.3 Genome replication

An understanding of the molecular mechanisms by which genomes are replicated occupies a central position not only in modern virology but in biology as a whole  for it is this process which ensures the continuity of the genetic information passed from one generation to the next.

It is common knowledge that the fidelity of nucleic acid replication stems from the fact that the base sequence along one strand of a duplex nucleic acid molecule is exactly complementary to that of the partner strand. Thus, given the sequence of one strand, the complementary strand may be constructed using the original strand as template. This holds for single- or double-stranded DNA and RNA genomes alike.

### 3.3.1 Replication of DNA genomes

The enzymology of DNA synthesis can be subdivided into the synthesis of the nucleoside triphosphates and their polymerisation into polynucleotide chains. Little will be said of the first aspect other than that some viruses depend entirely on the host cell metabolism for the supply of precursors, e.g phages $\phi$X174 and λ, while others such as T5, the T-even phages, herpesviruses and poxviruses induce enzymes for the synthesis of some DNA precursors. In many cases these virus-coded enzymes duplicate a function already present in the host cell.

Study of the enzymes essential for DNA syntheses has uncovered certain common groups of functions. These fall generally into four categories: polymerases, ligases, nucleases and 'unwindases'.

DNA polymerases extend polynucleotide chains by adding 5' nucleotides to the 3'-OH end, thus promoting chain growth in the 5' to 3' direction. Viruses such as T2, T4, T5 and T7 code for a DNA polymerase whose activity is essential for DNA replication; others such as phage λ do not specify their own polymerase and use the host-coded enzyme.

DNA ligases function by sealing single strand nicks in duplex DNA molecules by the formation of a phosphodiester bond between the 3'-OH end of one single strand and the adjacent 5'-phosphate end at the other side of the nick. For a double-stranded molecule with a gap in one strand the combined action of DNA polymerase, to fill the gap, and ligase to seal the nick, will completely repair the gap to form a continuous polynucleotide chain.

A variety of DNA-specific nucleases are involved in the replication process and for several phages, essential genes coding for nucleases have been identified. These nucleases include endonucleases which nick DNA chains and exonucleases which sequentially remove nucleotides from the ends of chains. Their specific functions in replication are not well understood but possible roles will be discussed below.

'Unwindase' describes a protein which takes its name from its postulated role in replication. This type of protein is believed to bind to replicating DNA just ahead of the growing point causing localized separation (unwinding) of the strands, thereby permitting synthesis of the nascent daughter strands to progress along the template. Two phage-coded essential functions having this property have been identified, the product of gene 32 of T4 and of gene 5 of the single-stranded DNA phage M13.

All known DNA polymerases catalyse DNA replication by adding nucleotides to the 3' end of a polynucleotide chain, that is, *all* DNA chain elongation proceeds in a 5' to 3' direction. In addition there is evidence to suggest that chain elongation commonly occurs by the synthesis of short fragments 1000–2000 bases long. These fragments, first identified by R. Okazaki, and referred to as 'Okazaki' fragments, are nascent structures which are rapidly joined by ligase action to the pre-existing chain. The problem of how the growing point of a replicating DNA duplex can progress along both template strands in the same overall direction (5' to 3' on one strand and 3' to 5' on the other) can be explained if 3' to 5' chain growth proceeds by a series of short steps each involving the synthesis of an 'Okazaki' fragment in the reverse direction from that of overall replication (Fig. 3.7).

The mechanisms by which viral DNA molecules are replicated is illustrated by reference to the Group I phages T7, λ and T4 and the Group II phage φX174.

T7 genomes exist as linear DNA duplexes with a unique base sequence and a terminal repetition accounting for about 0.6% of the molecular length. Viral DNA synthesis does not require the function of any known host DNA genes and is controlled by 7 T7 genes located at the left end of the genetic map. These T7 functions include a DNA polymerase, an RNA polymerase, a DNA ligase, an endonuclease and an exonuclease.

Replication of the parental DNA, and possibly all subsequent cycles of replication, is initiated at a fixed site 17% of the distance from one end of the parental genome. DNA synthesis progresses in both directions to produce initially an 'eye-form' molecule and latterly when the growing point on the shorter arm has reached the end a 'Y'-shaped structure (Fig. 3.8).

Soon after the start of DNA synthesis, intracellular T7 DNA can be extracted in the form of long linear molecules (concatenates) of up to four times the length of a viral genome. Concatenated DNA is produced at all subsequent times and serves as the precursor of

33

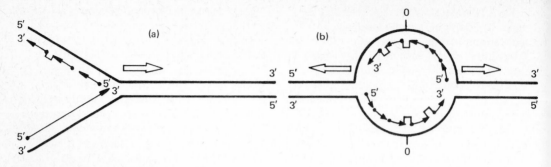

**Fig. 3.7** Schemes for DNA synthesis using Okazaki fragments. (a) Y-fork structure with continuous synthesis in $5' \rightarrow 3'$ direction and discontinuous synthesis in $3' \rightarrow 5'$ direction. (b) Eye-form structure replicating discontinuously in both directions from the internal origin (0). Thick lines indicate parental strands, thin lines denote newly-synthesized strands with arrows showing direction of chain extension, dots marking initiation sites and brackets indicating the ligase sealing of two fragments. By joining the two ends of structure (b) a theta-shaped molecule is produced.

progeny viral genomes. The mechanism by which concatenated molecules are produced is not completely understood, however, a scheme proposed by J.D. Watson, envisages the end-to-end joining of the linear daughter replicas followed by sealing of the joints. This proposal is supported by the observation that intracellular T7 molecules have short $3'$-terminated single-stranded ends; hybridization between the ends of different molecules would therefore produce concatenates. Prior to maturation the concatenates are converted to phage-equivalent lengths by a process requiring the function of several late genes, some of which are concerned with the synthesis and assembly of T7 phage heads. It appears likely that the activity of specific nucleases, which cut concatenated DNA at, or near the sequences corresponding to the natural ends of mature genomes, are involved.

DNA molecules extracted from phage λ, like those from T7, are linear DNA duplexes of unique sequence with short single-stranded complementary cohesive ends. Only three known viral functions are required for DNA synthesis; the products of gene N, which is required for delayed early gene transcription, and of the genes O and P. The activity of some of the host's DNA replication genes are also necessary.

Within 5 minutes after infection the linear parental genomes are converted by means of the cohesive ends and DNA ligase activity to circular molecules which are covalently-bonded throughout. During the next 15 minutes the parental circles replicate several times to produce about 20 circles per cell. Electron microscopic examination of these replicating circular forms show them to have two branch points, rather like the shape of the Greek letter theta ($\theta$) or an 'eye' form molecule with the two ends joined. Two of the three segments are of equal length and represent the duplicated segment; the length of one of these plus the third segment is equivalent to the total length of a single λ genome. Replication is initiated in all molecules at the same point and progresses bidirectionally from that origin. Comparison of λ and T7 early replicating forms reveals a marked similarity of structure at the replicating forks and clearly points to a basically similar mechanism. The fact that one is linear and the other is circular is in this respect of little significance, except to illustrate that at, or near, the growing points of a circular molecule there must exist single-strand breaks to permit rotation of the molecule as the parental duplex unwinds.

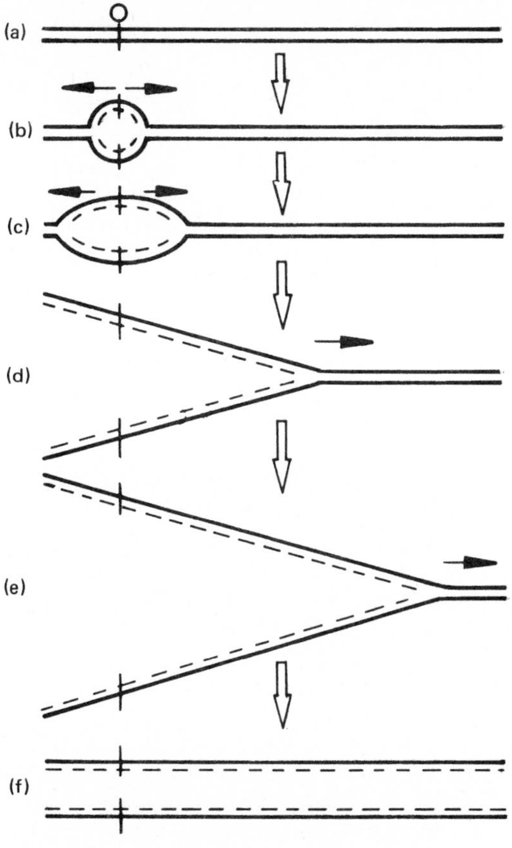

The intracellular synthesis of λ circles is superceded by the formation of concatenated structures with lengths of up to five times greater than the viral genome. Concatenate synthesis persists until lysis. Concatenates are derived from circular molecules and are the precursors of mature phage genomes.

There is good evidence that λ concatenates are synthesized by a rolling circle mechanism in which the circular replicating form is nicked and the 3′ end of the nicked strand acts as the primer for chain elongation using the closed strand as template. As the 3′ end extends, it displaces the 5′ end as a single strand which then becomes the template for synthesis of a complementary strand laid down as 'Okazaki' fragments to preserve the overall 5′ to 3′ direction of replication (Fig., 3.9.).

As with T7 the formation of unit length genomes from concatenates depends on the normal function of several head genes, the

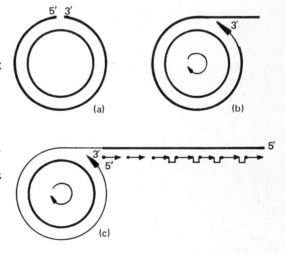

**Fig. 3.8** Bidirectional replication of T7 DNA (a) from a fixed origin (0) to produce initially an 'eye-form' structure (b) and (c) and then a 'Y-fork' structure (d) and (e). Newly-synthesized DNA shown as a broken line, horizontal arrows indicate overall direction of chain extension.

**Fig. 3.9** Rolling circle DNA replication. Parental and newly-synthesized strands shown as thick and thin lines respectively.

product of gene A being responsible for the formation of the cohesive ends.

T4 is an example of a phage with a permuted arrangement of sequences on its genome. As mentioned above, T4 is extremely virulent, causing chemical destruction of the host's DNA and showing a considerable degree of independence of host-coded functions. Normal DNA synthesis in T4-infected cells requires the function of about twenty genes all of which are virus-coded.

Concatenated DNA forms can be identified soon after replication. They contain both parental and newly-synthesized DNA and are the precursors for progeny phage genomes.

The association of head formation with the excision of mature length genomes from concatenates is also observed with T4. The mechanism of excision is unlikely to be sequence specific since the genomes consist of apparently randomly permuted sequences of constant length. It has been suggested that the length of DNA, cut from a concatenate by the maturation process, is determined by that length of DNA which can be packaged into a phage head. Excision of DNA headfuls, containing a complete set of gene sequences, plus a small terminal repetition would produce a collection of circularly-permuted sequences (Fig. 3.10).

Three stages can be defined during replication of the single-stranded circular genome of the Group II phage φX174. The parental genomes are first converted to double-stranded circular molecules by synthesis of the complementary strand. This duplex form is called the Replicative Form (RF). RF synthesis is catalysed by host cell enzymes and appears to be initiated at a unique site on each molecule and to proceed in a fixed sequence. RF molecules are then replicated several times by a semiconservative process until each cell contains about 30 RF molecules. Synthesis of daughter RF's requires the expression of a

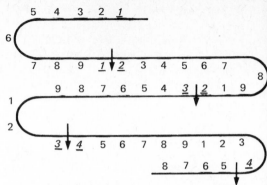

Fig. 3.10 Excision of DNA headfuls from a concatenate. A complete genome consists of sequences 1–9 (in any permutation) plus a small terminal redundancy. Arrows indicate sites at which the concatenate is cut; terminally redundant sequences are in italic.

phage-coded protein in addition to host DNA functions. This protein is the product of the A gene and is an endonuclease which nicks the viral strand of the RF at the specific origin of replication. The final stage of replication is the synthesis of single stranded circular genomes for maturation into virions. This event also requires A gene function and host cell enzymes, but in addition is dependent upon the activity of several genes which code for virion proteins.

There is evidence to suggest that all φX174 DNA synthesis occurs by a rolling circle mechanism differing somewhat from that of the general model presented above. The duplex tail formed during RF replication does not form a long linear molecule but folds back and by a recombination event produces a duplex circle. A new linear tail is then formed which repeats the process. During stage three of replication the synthesis of the complementary strand is suppressed.

Studies with the Group II phage M13 have linked RNA synthesis with DNA replication and similar findings have been reported for other viruses. It appears that DNA synthesis is primed by a short sequence of RNA laid down on the template. Furthermore, the synthesis of all 'Okazaki' fragments may be RNA primed, the RNA segment being subsequently displaced from the template. This reaction may therefore be of general significance for all types of DNA replication.

### 3.3.2 Replication of RNA genomes

RNA virus genomes unlike other RNA molecules serve as a template for their own replication. This means that a new type of enzyme, one capable of synthesizing RNA from an RNA template (RNA dependent RNA polymerase, or replicase) must be involved. The viral replicase of phage Qβ is a complex of several different subunits. The core replicase consists of four subunits, α, β, γ and δ. The β subunit is the phage-coded replicase protein and the remaining three are host proteins. The host proteins are not subunits of the host DNA dependent RNA polymerase but are proteins involved in protein synthesis. Similar subunits have been identified as part of the f2 phage RNA replicase complex.

RNA replication proceeds by way of complementary strand synthesis, the viral plus strand being used as template for the complementary minus strand, this in turn serving as template for plus strand synthesis. The same core enzyme of the replicase catalyzes both reactions although minus strand synthesis requires additional host-specified factors for initiation. Both plus and minus strands are replicated from the 3' end of the template, i.e. the new strand is laid down in the 5' to 3' direction (Fig. 3.11). The presence of the 3' terminus is mandatory for replication since very little *in vitro* synthesis occurs if the genome is fragmented. However, the 3' terminal fragments will only serve as template for complementary strand synthesis if the fragments are about half the normal length or greater. This suggests that a

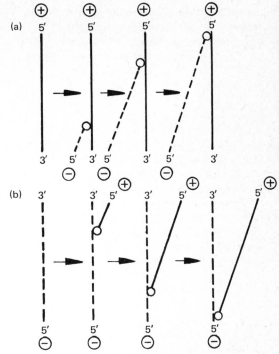

**Fig. 3.11** Replication of single-stranded RNA genome by alternating synthesis of minus strand on plus strand template (a), and vice versa to form progeny plus strands (b).

section in the middle of the genome is also required to initiate replication and has led to the idea that the replicase binds initially to this internal site. Because of the tertiary structure of the folded genome the replicase enzyme is placed adjacent to the 3' end of the molecule, to which it then binds and initiates replication. This model neatly sidesteps the problem that ribosomes and replicase molecules are both competing for a place on the genome and that they traverse the molecules in opposite directions. By binding at an internal site which also promotes ribosome binding the replicase prevents further ribosome binding and prevents its own displacement. When the last of the resident ribosomes have reached the 3' end of the genome the replicase can attach to begin replication.

The replication of animal virus RNA genomes is much less well understood than that of RNA bacteriophages, largely because replicase purification has proved to be extremely difficult. The poliovirus replicase, for example, is unstable *in vitro*, with low specific activity compared to that in infected cells, and has not been purified to a state of dependence on exogenous templates.

The single-stranded template complementary to the genome, which is produced during genome synthesis, may be − strands (Group IV) or + strands (Group IV). Genome replication of Group IV viruses (picorna- and togaviruses) takes place in membranous replicative complexes which contain RNA, proteins and replicase activity. The general mechanism of RNA synthesis is presumed to resemble that of bacteriophage Q$\beta$. Poliovirus RNA synthesis begins very shortly after infection and proceeds at an exponential rate until at least 10−20% of the total viral RNA has been synthesized. The rate of RNA synthesis then becomes linear. Other controls exercized over viral RNA replication include mechanisms which ensure the accumulation in each cell of optimal proportions of virion RNA and protein for virion assembly, and which allow the accumulation of an appreciable amount of intracellular viral RNA before virus assembly is initiated.

The synthesis and assembly of segmented genomes presents some interesting topological problems and no scheme has yet been proposed which unifies the mechanisms responsible for genetic reassortment and the production of virions containing a full complement of different genome segments. Reovirus genome segments appear to remain associated with each other throughout replication and nascent double-stranded RNA is associated with viron-like particles. It is possible that mRNAs may become associated with, and linked by, nascent viral proteins during, or shortly after, their synthesis, and that genome synthesis is completed inside newly assembled virus particles. The processes of transcription and replication are functionally distinct, but because of the association of both activities with virions and virion cores it has been suggested that replicase may be derived from transcriptase by the addition or modification of protein subunits. After the completion of genome synthesis and virus maturation, the replicase could then be converted back to transcriptase, ready to act in the next virus growth cycle.

## 3.4 Virus assembly

The simplest way in which a virus can be constructed from its components is by self assembly. The virions of many plant viruses, including those with both helical (TMV) and icosahedral symmetry, form in this way and can, in fact, be assembled *in vitro* by incubating a mixture of the virus coat protein and RNA at the correct pH, salt concentration and temperature. TMV is the best studied example and *in vitro* studies with this virus show that most of the information required for assembly is contained in the protein subunits, as alone they can form rod-like structures under appropriate conditions. TMV-RNA however, plays an important role in determining the length of the virus particle, as well as inducing the protein subunits to form helices rather than rods comprised of stacked discs.

Self-assembly also occurs in the final stages of formation of the virions of large DNA bacteriophages. The heads, tails and tail-fibres of these viruses are assembled separately in the infected cell and finally join together to form infectious phage particles. This final stage in virus assembly can occur *in vitro* when the preformed virus components are mixed and incubated under suitable conditions. However,

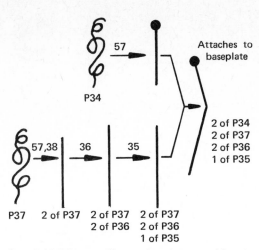

**Fig**. 3.12 Diagram illustrating the assembly of phage T4 tail fibres.

*in vitro* assembly of phage heads, tails and tail-fibres from their constituent proteins has not been achieved and there is good evidence that the formation of these structures involves non-structural catalytic components and host functions. The assembly and efficient attachment of the tail fibres of phage T4 to phage particles, for example, require at least 7 phage gene functions, whereas tail fibres themselves contain only four different polypeptides (Fig. 3.12). Two of the non-structural gene products are required for assembly of tail fibres (P38 and P57). Evidence from gene dosage experiments suggests that the function of P38 is catalytic and P57 stoichiometric; their exact functions are unknown. The product of gene 63 also has a catalytic function; it promotes the rate of fibre attachment to fibreless virus particles. The possible role of host functions during tail fibre construction is suggested by the existence of a *ts* mutant of *E. coli* which will not support the growth of phage T4 at 43° − infected cells at this temperature contain fibreless, but otherwise normal, phage particles, plus non-functional tail fibres. Other strains of

*E. coli* which will not support the growth of T4 phage have been found to affect the formation of phage heads.

Assembly is not regulated by mechanisms controlling the time of synthesis of components but is controlled by the specificity of component interactions. Thus the polypeptides required for tail fibre formation are all synthesized simultaneously, whereas assembly occurs in a series of sequential steps (Fig. 3.12). Defective polypeptides usually prevent the assembly process from going to completion (i.e. the production of tail fibres capable of interacting with phage heads and tails to give infectious virus) thus making the assembly of the virus an efficient process by ensuring that only normal tail fibres are used in the final stages of virus assembly. Another requirement for efficient phage assembly is that the necessary polypeptides have to be synthesized in the correct relative amounts; it has been shown that the relative underproduction of some structural components can seriously interfere with phage assembly. In general, the rates of synthesis of different structural polypeptides in the infected cell reflect the relative amounts of these polypeptides present in the phage particle. The assembly of phage heads and tails proceeds in a similar sequential fashion to that of phage tails, the products of genes coding for non-structural polypeptides being involved. Some structural polypeptides of the phage head possess shape-specifying properties which determine, for example, the formation of terminal caps or ends of the head and head size. Host factors, possibly membranes, are also involved in head assembly.

The details of DNA packaging into the head have not yet been fully worked out, but evidence is accumulating which favours the headful hypothesis, i.e. that head size determines DNA content. Aberrant phage particles containing abnormal structural proteins that give rise to small (petit) or large (lollipop) heads show interdependent head size and DNA content.

Without large numbers of genetically and biochemically characterized amber mutants our knowledge of phage assembly would still be rudimentary. The lack of such mutants in animal virus systems partly explains why relatively little is known about the assembly of these viruses, although intermediate stages in the assembly of some animal viruses have been demonstrated by biochemical techniques and by electron microscopy.

The structural proteins of animal DNA viruses with icosahedral symmetry are synthesized in the cytoplasm and then move into the nucleus where capsid assembly takes place. The details of these processes are obscure. It is not known, for example, if non-structural proteins are required and it is not known whether capsid formation precedes DNA packaging.

The icosahedral particles of herpesviruses acquire their outer envelope by budding through regions of nuclear membrane modified by the insertion of virus structural proteins. This process resembles the final stages of assembly of the enveloped RNA viruses.

An early stage in the assembly of the poliovirus virion is the formation of structural units from cleavage products VP0, VP1 and VP3. (Fig. 3.6) Five of these structural units combine to form another intermediate with a sedimentation value in the ultracentrifuge of 14S: twelve 14S particles then associate to form an 80S procapsid — an empty virus particle. RNA then associates with the procapsid and formation of the 150S virion is completed by the cleavage of polypeptide VP0 to polypeptides VP2 and VP4. The exact relationship between the entry of RNA into the procapsid and the cleavage of VP0 is not clear, and the enzyme responsible for the latter event has not been identified.

The assembly of enveloped RNA viruses occurs in three major steps. These are: (1) the

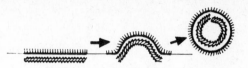

Fig. 3.13 Diagram showing the budding of an enveloped virus from the cell membrane.

assembly of a nucleocapsid; (2) the insertion of structural proteins into the host cell plasma membrane; and (3) a budding process where the virion nucleocapsid is enveloped by the modified plasma membrane and released from the cells (Fig. 3.13). Little is known about the assembly of virion nucleocapsids. As host proteins are not detectable in the membrane of mature virus particles, it would appear that virion membrane proteins either displace host membrane proteins during virus assembly, or that they attach to newly synthesized membrane. Many of the membrane proteins of the virion are glycosylated; host-glycosyltransferases are probably responsible for this.

### 3.5 Virus release

The growth cycle of most bacteriophages ends with lysis of the host cell and the subsequent release of mature infectious particles into the surrounding medium. The lysis of cells infected with bacteriophage T4 is triggered by hydrolysis of the mucopeptide layers of the cell wall by lysozyme, a phage coded enzyme. Lysozyme is synthesized and accumulates in the cell for a considerable time before lysis, and the timing of its action is dependent on phage induced and controlled alterations in the cell membrane, which allow the enzyme to reach and attack the cell wall.

The release of some animal viruses superficially resembles that of bacteriophages. Poliovirus-infected cells, for example, often rupture in an explosive fashion at the end of the virus growth cycle, releasing virus particles into the surrounding medium. The molecular basis of this process is not understood. The release of many animal viruses, does not, however, require cell lysis. Enveloped viruses, for example, are released from the cell surface by a budding process. This process obviously requires an intact cell. Some of these viruses can spread from cell to cell by causing the fusion of adjacent cell membranes (e.g. herpesviruses, poxviruses) and the virus particles can thus spread from cell to cell without exposure to the external environment.

## 3.6 Temperate phage and lysogeny

Our discussion of phage replication has been concerned mostly with events occurring during the lytic cycle of growth, in which the infecting phage directs the synthesis of a large number of progeny phages which are released by cell lysis. The lytic growth cycle is the only course open to virulent phages, e.g. T4, T7, $\phi$X174. However, many phages, probably the majority, can participate in an alternative interaction with their host in which the phage and host multiply together. This relationship is called lysogeny and phages capable of this type of interaction are known as temperate phages. The form of phage present in lysogenic cells is called prophage.

A susceptible cell infected with a temperate phage may enter the lytic or a lysogenic cycle depending on a variety of genetic and physiological conditions. Once the infecting phage has become prophage, the lysogenic cell may multiply for many generations. Lysogenic bacteria have two main properties: (1) they are capable of producing infectious phage i.e. the prophage may become vegetative phage, replicate and lyse the host cell; and (2) they cannot undergo a lytic infection with the same or a closely related phage (superinfection immunity).

Much of the interest in lysogeny revolves about two problems; what is the form of prophage and what control mechanisms operate to regulate the transition between lysis and lysogeny?

### 3.6.1 The nature of prophage

The fact that lysogenic cells carry the capacity to release normal infectious phage particles means that all cells must harbour a complete copy of the genetic information necessary for phage replication. Artificial lysis of lysogenic cells does not, however, liberate infectious phage particles and so prophage is not synonymous with the virion. The number of prophage genomes per lysogenic cell is small and for phage $\lambda$ is approximately equivalent to the number of bacterial chromosomes. Moreover, it is difficult to lysogenize a cell with more than one prophage of a given type. Lysogenic cells cannot therefore contain vast numbers of prophage which become distributed among many daughter cells by dilution of the original set of copies. More likely the prophage replicates in parallel with the cell chromosome and it is the daughter prophage molecules which are transmitted to the daughter lysogenic bacteria. Proof of this idea came from studies of chromosome transfer between strains of *E. coli*. Briefly, the conjugal transfer of DNA from a donor to a recipient cell occurs in a fixed sequence corresponding to the order of genes on the chromosome. The origin and direction of transfer can vary from one donor strain to another but for a particular donor strain the order in which the donor's genetic markers are transferred into the recipient cell is unique. Crosses between a donor lysogenic for phage $\lambda$ and a non-lysogenic recipient showed that in addition to the transfer of bacterial markers, the $\lambda$ prophage could also be transferred to the recipient. Moreover, the prophage was inherited by the recipient in the manner of a bacterial marker, i.e. its transfer occurred in a fixed sequence relative to that of the donor markers. This type of study provides evidence that the prophage is associated physically with the bacterial chromosome at a specific site. In general, other prophages behave similarly except that each has a different and unique chromosomal location. Exceptions to this rule are known, e.g. phage Mu-I can integrate at any site on the *E coli* chromosome (over 70 different sites have been identified within a single gene) and phage P1 appears not to integrate. The DNA of host and prophage genomes are contiguous as shown by the decrease of genetic linkage of bacterial markers on either side of the prophage and from the fact that the deletion of DNA from one end of prophage can lead to deletion of adjacent host genes. The gene order on the λ prophage is identical to that of the vegetative phage. However, while both are unique, one is a permutation of the other. This finding prompted A.M. Campbell to suggest a mechanism for prophage insertion in which the λ genome first becomes circular and then integrates into the bacterial chromosome by a single reciprocal recombination event between specific sites on phage and host genomes (Fig. 3.14). The site on the phage chromosome at which this recombination occurs is termed *att* (for attachment) and the process is controlled by the specific recombination function of the λ *int* (integration) gene. Surprisingly this recombination is not determined by base sequence homology between the two chromosomes.

### 3.6.2 The control of lysogeny

As mentioned above, not every cell infected by a temperate phage becomes lysogenic, a proportion of infections proceeding to the lytic cycle. Equally, the lysogenic state is not permanent. At a low frequency the lysogenic state is converted to the lytic state, with subsequent vegetative replication and cell lysis. This spontaneous breakdown of lysogeny can be greatly enhanced (up to 100%) under certain

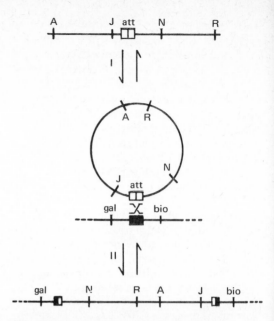

**Fig. 3.14** Integration of λ genome into a chromosome of *E. coli*. Circularization of genome (Step I) is followed by a recombinational event involving the prophage attachment site (att, □□) and equivalent site on bacterial chromosome (■■), (Step II). Prophage excision is the reverse process.

conditions – a process known as induction. Exposure of lysogenic cells to UV or the DNA synthesis inhibitor mitomycin C leads to induction. Mutations in several phage genes affect lysogeny, one class leading to the formation of clear plaques instead of the usual turbid plaques. (Turbid plaques result from the formation of some lysogenic cells which by virtue of their superinfection immunity are not lysed by the neighbouring phage.) Clear plaque mutants (*c*) have been studied extensively for λ and P22, with similar results. Genetic analysis of λ *c* mutants identifies three complementation groups which map close together, *cI*, *cII* and *cIII*. All three are required for lysogeny, *c*I for maintenance and *c*II and *c*III for initiation. By using genetic tricks it is possible to promote integration of *c*II or *c*III mutants to form stable lysogens carrying *c*II or *c*III mutants as prophage – thus the continued function of the *c*II and *c*III genes is not essential for maintaining lysogeny. This is not so for *c*I mutants. Moreover *c*I mutants can be obtained which produce a temperature-sensitive product and while lysogeny is maintained normally at a low temperature, elevation of the incubation temperature leads to simultaneous induction and loss of immunity. Therefore the *c*I product also governs both prophage maintenance and immunity. Another facet of lysogenic control is illustrated by the results of bacterial matings in which λ prophage is transferred by a donor into an non-lysogenic recipient – entry of the λ prophage into the λ-sensitive cell causes induction. This is known as zygotic induction. Studies of this type produced the idea that lysogeny (prophage maintenance) and immunity are both under the control of the *c*I gene product which was considered to be a diffusible protein (to account for superinfection immunity) which functions as a repressor to block the expression of vegetative functions. Repression of the vegetative expression of the phage genome maintains the genome in the prophage state and also confers immunity to the lysogen by preventing the vegetative expression of superinfecting genomes. Induction results from destruction of the repressor and as a consequence leads to expression of the phage functions required for prophage excision and replication. The specificity of immunity is determined by specific sites on the phage genome at which the repressor molecule acts. This model has been fully substantiated.

Isolation of the *c*I gene product revealed it to be a protein which binds to λ DNA specifically at operator sites adjacent to and on either side of the *c*I gene. The location of the operator sites places them near to the promoter sites for early gene transcription. Mutations in the operator regions which reduce their affinity for repressor, permit the phage to replicate in lysogenic cells, i.e. they are not subject to immunity.

Therefore we see a pattern of negative control where repressor synthesis leads to the inhibition of transcription of segments (operons) of the λ genome which flank the immunity region and which include the recombination and prophage excision genes (leftwards transcription) and the genes required for genome replication (rightward transcription). Blocking these two sites is sufficient to keep the prophage integrated and to prevent its replication should it be excised.

Another control is exerted by the product of the *cro* gene. This product is termed the repressor-regulator since it can prevent *c*I repressor synthesis. However, in immune lysogens the presence of *c*I repressor, blocks *cro* synthesis. *Cro* function also plays a role in the lytic cycle by shutting off early transcription of the leftward and probably the rightward operons.

Upon infection of a sensitive cell, proteins required for vegetative (lytic) growth and for lysogeny are both synthesized before the system is committed to one of the two processes. It has been suggested that the ratio of concentrations of the two regulatory proteins *c*I and *cro* may determine the outcome since *c*I and *cro* products block the lytic and lysogenic pathways respectively. Moreover, *c*I product inhibits *cro* synthesis and vice versa so that once the ratio moves in a particular direction there is a cascade effect which completely topples the balance in that direction.

# 4 Virus genetics

Phage systems have been the subject of intensive genetic analysis during the past 30 years. This work provides virtually the entire basis for our understanding of the genetic properties of viruses and moreover has contributed extensively to the formulation of more general genetic principles. The genetic analysis of animal viruses has been the subject of rigorous study only within the last five years and for plant viruses little data is available. The reasons for this are partly historical, but largely technical — phage systems being more amenable to study than plant and animal viruses. Consequently, most of our discussion of virus genetics will rely on information obtained with phage systems.

As with other biological systems the genetic analysis of viruses requires a collection of genetically variant individuals (mutants) and a suitable means of permitting genetic interactions between these mutants. The development of phage genetics really dates from the mid-forties with the isolation by A D. Hershey of plaque morphology mutants of phage T2. This observation was followed by the demonstration that more than one phage could replicate within the same bacterial cell and that the progeny phage released from single cells infected with a mixture of wild type and mutant phage contained both phage types. It was a short step to infect cells with a mixture of two parental phages differing by two mutations and reveal that the progeny contained not only the two parental types but two recombinant types bearing the parental mutations in the alternative combinations. Therefore the virus-infected cell constitutes the mating environment within which genetic interactions leading to recombinant formation can occur.

## 4.1 Mutations

Many types of mutation are available in viruses, not all of which are understood in functional terms and not all of which are suitable for use in genetic studies. Mutations arise from alterations of the nucleotide sequence of the genetic material. The effect of this alteration on the phenotype of the organism will vary. At one extreme there may be no effect, e.g. the substitution of one base by another may have no noticeable effect if the new codon specifies the same amino acid as the wild type codon. Conversely, the mutation may involve the loss of a series of nucleotides in a gene coding for a polypeptide whose normal function is vital for replication. Such a mutation would be lethal and would not normally be identified. Mutations useful for genetic studies either fall between these extremes, i.e. severe enough to cause a distinct phenotypic difference but not so severe as to prevent propagation, or are lethal under some conditions but not under others (conditional lethal mutations).

### 4.1.1. Specific mutations

Specific mutations can be considered as those causing a specific phenotypic effect limited to a particular gene or a small group of genes. They can be very useful for genetic analysis particularly if they can be scored by uncomplicated methods such as direct observation of plaques on a plate or by selective methods which distinguish mutant and wild type alleles. The phenotypic effect will vary according to the specific mutation, e.g. mutations which render T4 phage particles resistant to inactivation by osmotic shock are located in the gene which specifies the head protein subunit, the production of clear plaques in phage λ results from mutations in three genes, $cI$, $cII$ and $cIII$. Plaque morphology mutants have been of paramount importance in the genetic analysis of viruses, the rapid lysis ($r$) mutants of the T-even phages being notable in this respect. First identified and isolated by A. D. Hershey, the $r$ mutants produce plaques on $E. coli$ strain B, having a clear-cut perimeter quite distinct from the fuzzy edge of wild type ($r^+$) plaques. Host range mutations which affect the range of host strains capable of supporting replication of a given virus have also been of great value. Specific mutations of various kinds have been isolated from many bacterial and animal viruses.

### 4.1.2 General systems

Virus genetics and molecular biology was advanced in a dramatic and lasting way by the identification in the early 1960's of two types of general mutation. These mutations are conditional lethal mutations, which, as their name implies, are lethal under some conditions and viable under others. Their particular importance, and the reason for considering them as general mutational systems is that they can occur in virtually all viral genes and so can affect functions essential for virus replication. From a genetic standpoint conditional lethal mutations provide almost limitless numbers of mutant sites, spanning nearly the entire genome. For biochemical studies they can provide a complete collection of defective viral functions.

The two major classes of conditional lethal mutations available are: (a) temperature-restricted mutants, and (b) nonsense or suppressible mutants. Temperature-restricted mutants replicate successfully within a narrower temperature range than wild type virus. A wild type phage of $E. coli$ will replicate efficiently between about 30°C and 40°C; above this temperature the cells stop growing and at lower temperatures the replication rate is reduced to a practically undesirable level. (Other virus-cell systems may have different temperature ranges, e.g. for viruses infecting amphibian cells the temperature range will be much lower.) The most commonly used type of temperature-restricted mutant is the temperature-sensitive ($ts$) mutant which is sensitive to the upper range of temperatures, i.e. the mutant will grow more or less normally at 30°C (the permissive temperature) but fails to replicate at 40°C (the nonpermissive temperature).

$ts$ mutants are missense point mutations, that is, one base of one coding triplet is altered by a single base substitution, resulting in the replacement of one amino acid by another in the polypeptide gene product. Direct evidence is available which reveals that the function of the mutant polypeptide can be more temperature-sensitive than the wild type protein and this is probably the most common basis of the mutation. $ts$ mutations are ubiquitous and have been isolated wherever sought regardless of the type of virus.

To date, nonsense or suppressible virus mutations have only been isolated from phage — both DNA and RNA-containing. They were first isolated by A.M. Campbell in phage λ as mutants which were lethal in some bacterial strains but could replicate satisfactorily in others. Bacterial strains of the latter group were defined as suppressor strains, being able to suppress the lethal mutant phenotype, and the phage mutants were named $sus$ mutants. At about the same time S. Benzer noted that some rII mutants (one class of r mutants) of phage T4 were ambivalent, being unable to replicate on one strain of *E. coli* K but active on another. Finally, R.S. Edgar, R.H. Epstein and their colleagues isolated a series of T4 mutants unable to replicate on the normal T4 host, *E. coli* strain B, but able to replicate on a K12 strain. These mutants occurred in many different genes and were named amber (*am*) mutants. With characteristic foresight Benzer had suggested that the rII ambivalent mutants might be nonsense mutants and that the bacterial strains permitting growth of the ambivalent mutants were suppressing the lethal phenotype by allowing the nonsense to be read as sense. In fact, *sus*, ambivalent and *am* mutants are now all known to be nonsense mutations in which the mutation has converted a sense triplet, coding for an amino acid, into a nonsense triplet for which no amino acid is available. The effect of the mutation is evident during translation of the mRNA when the translation process is halted at the site of the nonsense mutation and the incomplete poly-peptide is prematurely released from the ribosomal complex. Three of the 64 RNA triplets of the genetic code, UAG, UAA and UGA, do not specify an amino acid and constitute nonsense triplets. All amber mutants have the triplet UAG at the mutant site. Nonsense mutations containing the UAA and UGA codons are also known. Suppression of nonsense mutations is a property of the bacterial transfer RNA, suppressor strains carry a mutation in a minor tRNA species which leads to recognition of a nonsense as a sense codon and inserts an amino acid. Several types of mutant transfer RNA have been identified from suppressor strains of bacteria, each inserting a different amino acid in the growing polypeptide chain at the site of the nonsense codon. This prevents termination of polypeptide synthesis at the nonsense triplet and produces a full-size protein capable of nearly normal function.

## 4.2 Functional or complementation analysis

Complementation is the process whereby two mutant genomes when present in the same cellular environment can supply each others deficiency to produce a wild type or near wild type phenotype. For viruses, complementation is observed when the two mutants infect the same host cell. The most efficient and common form of complementation occurs when the mutations affect different functions — intergenic comple-mentation. A much rarer type of complementa-tion can occur between mutants located within the same gene — intragenic complementation.

Complementation is ideally illustrated with conditional lethal mutations where it has been used extensively as a method to classify the mutants into functional groups (cistrons or genes). Consider three different amber phage mutants, two mutants, *am*1 and *am*2, located in one gene and the third, *am*3, in a different gene (Fig. 4.1). Non-permissive cells are infected with the mutants singly and in the three pairwise combinations. The single infections will by definition be non-productive. Cells infected with *am*1 and *am*2 will also be non-productive because the mixedly infected cells are still defi-cient for an active (wild type) gene product specified by the mutant gene. Infections with mixtures of *am*3, or *am*2 and *am*3 will, how-ever, be productive since the infected cells con-tain a complete collection of all gene products specified by the wild type virus, the faulty product of one mutant parent being con-tributed by the other. The mixedly infected cell is phenotypically wild type and will produce a normal or near normal crop of progeny virus. The progeny virus will still be genotypically mutant and like their parents will be non-productive in further single infections of the non-permissive host, (recombination between the mutant parents might occur and so a pro-portion of the mixed infection progeny could be recombinant).

Theoretically the complementation test is

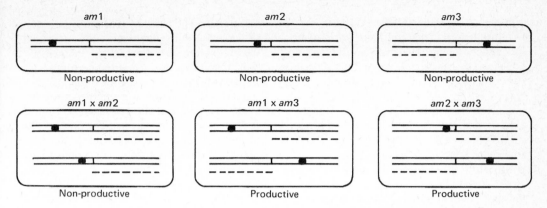

**Fig. 4.1** Genetic complementation. Mutations *am*1 and *am*2 marked by a black dot are in the same gene and therefore are non-productive in mixed infection. When either is crossed with *am*3, the infected cell produces a complete set of wild type gene products (dotted line) and consequently a productive infection ensues.

important since it defines unity of function of genes and its use has been instrumental in clarifying the concept of the gene. Its practical use is equally important, the geneticist, having isolated say a 100 conditional lethal virus mutants, all with the same basic phenotype — inability to replicate under a standard set of conditions — can set about classifying this collection into different genes on the basis that mutations within the same gene do not complement each other. The rare occurence of intragenic complementation does not invalidate the usefulness of this type of classification.

### 4.3 Fundamental studies in recombination
By the mid-fifties a collection of fundamental observations had accumulated to the point where a general description of phage recombination could be attempted. Before considering this data we should be familiar with the procedure for making a phage cross. A standard cross is performed by infecting a population of about $10^8$, logarithmically-growing, sensitive bacteria with a mixture of parent phages. The phage mixture will contain equal numbers of both parental phages in sufficient quantity

such that each cell is infected with an average of about 10 particles. Following a brief period for phage adsorption, the infected cells are diluted into growth medium and incubated for a time sufficient to allow all cells to lyse spontaneously, i.e. one latent period. The progeny phage are then analysed.

As mentioned above, a 2-factor cross in which the parents differ at two mutational sites $a\,b^+$ × $a^+b$, will produce progeny with four genotypes, the two parents $a\,b^+$ and $a^+b$ and two recombinants $a\,b$ and $a^+b^+$. For a given pair of mutants the percentage of recombinants among the progeny from a standard cross is constant with the two recombinants being produced in equivalent numbers. The recombination frequency is independent of the parental arrangement of markers i.e. $a\,b^+$ × $a^+b$ or $a\,b$ × $a^+b^+$. For different pairs of markers the recombination frequency varies in a manner consistent with the notion of genetic linkage such that mutational sites can be ordered to form a genetic map which is based on the additivity of recombination frequencies (map distances).

The foregoing description of phage recombination is consistent with that observed

47

with any of the well-studied eukaryotic systems, such as the mould *Neurospora* or the fruit fly *Drosophila*. However, several lines of evidence suggested that a phage cross cannot be viewed in quite the same way as with these organisms:

(1) The earliest maturing progeny from a cross contain fewer recombinants than populations obtained by spontaneous lysis. If lysis is artificially delayed the recombinant frequency is higher still. For loosely linked markers the recombination frequency reaches about 40%. This suggests that phage chromosomes are involved in repeated matings, resulting in a drift towards genetic equilibrium.

(2) Bacteria infected with three parental types will release a proportion of recombinants combining markers derived from all three parents. This result also suggests multiple matings.

(3) Examination of the cross progeny from individual bacteria reveals little or no correlation between the numbers of opposite type recombinants.

(4) In a two-factor cross with unequal input of parents the progeny may contain more recombinants than particles with the genotype of the minority parent.

These observations can be interpreted in terms of a theory advanced by N. Visconti and M. Delbrück (1953) to explain the recombination properties of phages T2 and T4. The theory considers that a phage cross must be viewed as a problem in population genetics. Upon entering a sensitive bacterium the parental phages replicate to form a mating pool (the DNA pool) in which the genomes undergo repeated pairwise matings which are random for partner and with respect to time. Maturation terminates the mating experience of a given vegetative phage genome and once matured the phage neither mates nor replicates. The mating pool of replicating genomes is continually being replenished by replication and depleted by maturation. The result of repeated matings is seen as a rise in recombinants with time and by the production of triparental recombinants. The randomness of the maturation process would influence the frequency of opposite (reciprocal) recombinant genotypes from single bursts and their statistical equivalence from the content of large populations. The randomness of the mating process could result in the reduction of the minority parent to a proportion less than that of the recombinant types since it is very probable that early in infection its genotype will be lost by a mating with a genome of opposite genotype (with an equal input cross many of the early matings will be between genomes of the same genotype).

## 4.4 Chromosome mapping

The object of chromosome mapping is to derive a picture of the spatial relationship of genetic markets. The laws of genetics maintain that the genetic maps are related to chromosomes — unlinked markers occupy positions on different chromosomes, linked markers reside on the same chromosome — and that the spatial relationship of markers on the genetic map is related to the locations of the mutations on the chromosome. Viruses occupy a useful position in this respect, their chromosomes can be isolated intact as unadorned nucleic acid and in many instances are still infectious. In a simple case therefore, the genetic map would be linear and the arrangement of markers would bear a faithful scale representation of the mutational sites on the linear nucleic acid molecule. The construction of genetic maps also provides an important basis for understanding the genetic properties of viruses. For example, unusual aspects of the map may reflect atypical physical characteristics of the chromosome or variations in the recombination process. A comparison of the map location and functional properties of mutants might identify functional groupings important in the regulation of gene function (see Chapter 3).

Genetic maps are traditionally constructed by assembling linkage data obtained from crosses (linkage analysis). More recently, techniques have been developed which permit the construction of physical maps based on the arrangement of nucleic acid sequences on the genome rather than on recombination frequencies between mutants. This has permitted the revealing exercise of comparing genetic and physical maps.

### 4.4.1 Genetic maps

The principles governing the construction of genetic maps are basically the same for viruses as for eukaryotic organisms. Recombination frequencies obtained from 2-factor crosses involving a series of mutants crossed in all pairwise combinations can usually be arranged in a sequence based on the additivity of the recombination frequencies. Additivity of map distances is usually sufficient to locate mutations unambiguously but is rarely exact, and loss of additivity increases with the distance between mutants. Three-factor cross analysis will assist the ordering of markers and is not so dependent on precise measurement of recombination frequencies.

With the possible exception of some RNA viruses all virus mutations map in a manner consistent with each having a single linkage group. However, the maps are not always directly related to genomes as in our simple case mentioned above. Two examples will illustrate this point. (1) The genetic maps of T2 and T4 are circular. It can now be appreciated that a collection of linear genomes bearing gene sequences in a series of circular permutations can give rise to a circular genetic map. The temperate phage P22 shows similar properties. Other viruses have circular genetic maps by virtue of having circular genomes, e.g. phages $\phi$X174 and S13. (2) The genetic map of phage T1 is distinctly non-random with most of the markers mapping within the central region. Evidence suggests that genetic exchanges occur most frequently near the ends of the unique chromosomes. This effect would stretch map distances at the ends of the map.

Genetic recombination has also been identified in many animal virus systems. Genetic mapping, however, has not yet developed into a routine method and genetic maps of any size and with reasonably additive map distances are avaiblable for only a few, e.g. poliovirus, influenza virus, adenovirus type 5, herpes simplex virus type 1, pseudorabies virus. A genetic map has yet to be constructed for a plant virus.

RNA viruses pose special and interesting problems. In the first place no recombination has been observed for RNA phages but has been reported for several animal viruses. This raises the possibility of a real difference between phage and animal viruses. Secondly, several RNA animal viruses have fragmented genomes. This fact introduces the potential for the independent assortment of markers by the random withdrawal of sets of RNA fragments from the replicating pool of RNA genomes. Indeed for reovirus and influenza virus, recombination frequencies for pairs of mutants are either very low, suggesting that they reside on the same linkage group, or high, indicating the random assortment of genome fragments.

### 4.4.2 Physical maps and their relationship to genetic maps

It was previously stated that the arrangement of mutations on a genetic map is related to the sequence of the mutant bases on the nucleic acid molecule. The proof rests on studies which map chromosomes by physical rather than genetic methods. A good example is provided by phage $\lambda$ which has two useful properties; the extracted DNA is infectious when applied to cells and will recombine with $\lambda$ DNA molecules which have entered the cell by infection with intact phage particles (helper phage), and the left and right halves of broken λ DNA molecule can be separated. Mixed infections with purified left halves of wild type DNA and phage particles bearing suitable genetic markers produce recombinants involving markers only to one side of the genetic map. The same result is obtained with purified right halves, except that only markers on the other side of the map are represented on the DNA fragments. In an extension of this work the right half molecules were broken further and separated according to size. Each size class was tested for the presence of markers on fragments extending inwards from the right end. The result showed clearly that the shortest fragments carried only the R marker proximal to the right end of the genetic map and that increase in fragment length was accompanied by the sequential appearance of markers in their precise sequence on the genetic map (Fig. 4.2).

Various methods, particularly those involving the hybridization of specific nucleic acid sequences to single strands from intact genomes, have been used to establish a general equivalence between the positions of genes on maps and on chromosomes. In a few instances it has been possible to examine this congruence at a much finer level, within a gene in fact. As discussed above, in the absence of suppression, nonsense mutations cause termination of peptide synthesis at the site of the nonsense codon. The length of polypeptide synthesized will depend on the distance of the nonsense triplet from the end of the mRNA chain from which translation starts. Each of a series of amber mutants mapping at different sites within a gene should produce a discrete polypeptide with a length corresponding to the position of the nonsense triplet on the mRNA molecule. The molecular weight of each polypeptide can be precisely estimated and compared to the map location of the mutant obtained by linkage studies. Such an analysis

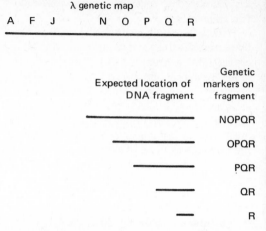

Fig. 4.2 Distribution of genetic markers carried on specific fragments of λ DNA molecules. Fragments were obtained by shearing right half λ molecules and fractionating into specific size classes.

was carried out by A. Sarabhai and S. Brenner for the gene 23 protein of phage T4, the major head protein. The genetic and chemical comparison revealed a very close agreement between map position and length of polypeptide fragment. This result provides excellent proof of validity of the genetic map as a reflection of the structure of the chromosome by showing that in this case the recombination frequency is constant for the length of the gene 23 base sequence.

Variation in the recombination frequency over a given length of chromosome will lead to loss of equivalence between map position and chromosome location. Several cases of recombinational variability along a section of chromosome have been reported. Thus while overall correspondence of the gene sequence on the linkage map and genome is the rule, close inspection can uncover regions of variable recombining ability.

## 4.5 Genetic recombination

The molecular basis of genetic recombination, i.e. the sequences of events by which virus genomes interact to form recombinant genotypes, is still rather obscure. However, considerable data is available and it is possible to create some reasonable, if not complete, pictures. The initial studies were entirely by genetic analysis of the genotypes produced from crosses. This approach attempts to predict the sequence of events by analysis of the products and has proved very fruitful. Latterly, with the introduction of sophisticated physical, chemical and biochemical techniques it has become possible to study the molecular structure of recombinant genomes and the enzymology of their production. Added to this has been the availability of large numbers of conditional lethal mutants, which have been used singly and in concert to detect effects on the normal recombination process.

One of the first important observations came in 1951 from a study by A.D. Hershey and M. Chase of the progeny from crosses between T2$r$ and T2$r^+$ phage. While most of the progeny phage produced either pure $r$ or pure $r^+$ plaques, about 2%, produced mottled plaques whose morphology was partly $r$ and partly $r^+$. These mottled plaques contain predominantly $r$ and $r^+$ phage and arise from single phage particles which are heterozygous for the two $r$ alleles. This heterozygosity is transitory since replication leads to segregation with the production of predominantly pure $r$ and pure $r^+$ descendents. For any pair of alleles, heterozygous particles are formed at this same low constant frequency. (Mottled plaques only arise from particles heterozygous for the $r/r^+$ alleles, or some equally obvious plaque morphology character — for other mutations identification is more difficult but their characteristics are the same.) For any phage particle the mean length of the heterozygous region of the chromosome is short, and particles heterozygous for one pair of

alleles will be heterozygous for other pairs of alleles only if they are very closely linked. Furthermore, among the progeny from a 3-factor cross the majority of particles heterozygous for the central pair of alleles are recombinant for the outside markers. These observations taken with the fact that the T2 DNA molecule is constructed from two continuous polynucleotide chains led to the conclusion that the likely structure for a heterozygous molecule is a DNA duplex with a short heteroduplex section at the site of the genetic heterozygosity (Fig. 4.3b). That is, the heterozygosity is a function of base sequence of the two strands. (A second type of heterozygote has been identified, but is not relevant to this discussion.) The formation of the heteroduplex is considered to be part of the recombination act; its segregation, by replication, completes the process. During virus replication the formation and segregation of heterozygotes occurs continually as a result of repeated matings and by chance a few are matured prior to segregation, to be released at lysis.

Many of the observations described for T2 have been repeated for other phages and similar evidence is available for herpes simplex virus. It appears probable that the partial heteroduplex heterozygote is a standard intermediate in recombinant formation.

The basic mechanism of recombinant formation was revealed by M. Meselson in a series of physical studies with phage λ. Two genetically marked parents, both labelled in the DNA with heavy density isotopes, were crossed in unlabelled cells and the progeny phage analysed in a density gradient. Some recombinants were found which were virtually completely heavy, that is, they had arisen by the breakage and reassociation of parts of parental genomes. Therefore recombination can occur without extensive DNA synthesis. In fact a small amount of DNA replication would have missed detection, moreover later studies by F.W. Stahl have clearly shown that many of the recombination events in λ are accompanied by measurable DNA synthesis. The point is, however, that recombination can occur by the breakage and reunion of unreplicated genomes.

Further evidence that recombinant formation can proceed by a breakage-reunion mechanism by way of heteroduplex formation comes from a study by J. Tomizawa of the structure of recombinant DNA produced from crosses with phage T4. P hage with heavy density-labelled DNA were crossed with phage containing radioactive DNA and at intervals after infection the intracellular DNA was extracted and analysed on density gradients. T4 DNA molecules could be identified which contained within the same structure both radioactive and density labels, i.e. were composed of segments from both parental molecules. At early times this association of the parental segments was unstable, being destroyed by conditions which disrupt hydrogen bonds. At later times this association was stable to denaturation. Thus recombination appears to result from formation of hybrid molecules bearing a heteroduplex overlap segment in which initially the ends of the overlap are not joined (joint molecules), at later times the ends become covalently linked to form recombinant molecules (Fig. 4.3).

A further important feature of the recombination mechanism is illustrated by the results of experiments designed to analyse the genetic products from single recombination events. In a study using phage fl, T. Boon and N. Zinder observed that among the progeny of single cells which had undergone one recombination event, the majority of cells yielded only one recombinant and one parental genotype. In other words, in addition to any effects on recombinant frequencies imposed by the random maturation of genomes, it also seems likely that the recombination event itself does not produce reciprocal recombinants.

Some understanding of the biochemistry of

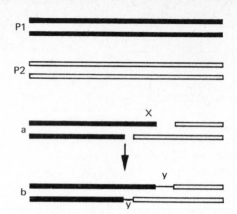

Fig. 4.3 Structures of intermediates in recombinant formation. Parental genomes P1 and P2 give rise to 'joint molecules' (a), held together by hydrogen-bonding in the hybrid region x. The action of DNA polymerase and DNA ligase at sites y will convert joint molecules to recombinant or heteroduplex molecules. Duplication of b will cause segregation of the heterozygote.

recombinant formation has resulted from the identification of mutations which affect normal phage recombination. Mutations of phage λ are known which reduce the normal recombination frequency by about 10-fold in wild type host cells. The mutations map in two adjacent genes, *redα* and *redβ*, the first of which codes for an exonuclease. When *red* mutant phage are crossed in recombination defective host cells the production of phage recombinations is essentially abolished. Therefore, in the absence of the phage recombination functions the host's recombination facility can be used. Mutations in several T4 genes reduce the recombination frequency and several code for functions essential for DNA synthesis. For example, the unwindase product of gene 32 is required to form the joint molecules. Conversion of joint molecules to recombinant molecules is catalysed by the T4 DNA ligase. Mutations in genes 46 and 47, which specify an endonuclease essential for

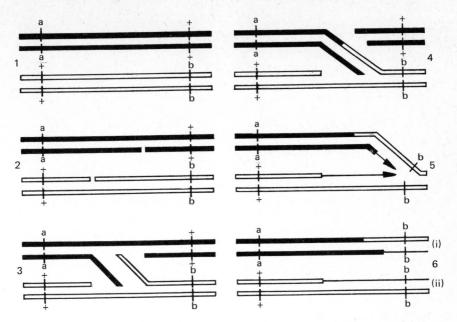

**Fig. 4.4** Recombination scheme proposed by Boon and Zinder. Parental genomes (1) are randomly nicked (2) and a heteroduplex is formed (3). Further breaks and joins (4) followed by DNA synthesis (5) leads to formation of one recombinant (6i) and one molecule of parental genotype (6ii). This sequence depicts the simplest version of the model.

continued DNA synthesis, also cause a significant depression in the recombination frequency.

In summary, current evidence points to a recombination mechanism in which DNA segments are recombined via a heteroduplex intermediate structure. Heteroduplex formation results from the nucleolytically-induced breakage of parental single strands followed by their reassociation with heterologous complementary strands. Some DNA synthesis (repair synthesis) will be necessary to fill any resulting gaps and discontinuities (Fig. 4.3). Recombinants are produced in a non-reciprocal manner. More extensive DNA synthesis, of the kind observed by F.W. Stahl, may also be an integral step in heteroduplex formation. Several models have been proposed to account for these observations, one of which is illustrated in Fig. 4.4. This scheme may not be entirely correct but serves to illustrate some of the molecular interactions which are considered to be part of the recombination process.

# 5 Tumour virology

## 5.1 What is a tumour?

In its original sense the word tumour described any abnormal swelling in an individual, but its use is now restricted to pathological conditions caused by the excessive and apparently purposeless growth of cells — often with harmful effects. Tumours may arise from almost any cell in the body — an interesting exception being the adult neurone — and often show a resemblance to normal types of cell and tissue. Tumours vary widely in their rates of growth and in the ability of their cells to invade surrounding tissues and to spread via the blood stream and lymphatics to form distant independent growths — metastases. The cells of benign tumours remain localized at their site of origin and are usually slow-growing, closely resembling normal cells in most respects. Malignant tumours contain cells with pronounced invasive properties and with the capability of producing metastases. Common types of malignant tumours include the carcinomas (tumours of epithelial tissues) and the sarcomas (tumours of connective tissue, cartilage and bone).

## 5.2 Causation of tumours

The molecular basis of tumour formation is not understood, although a long list of agents which cause tumours can be seen in any textbook on the subject. These agents, which are called carcinogens, include chemicals, radiation and viruses. Peyton Rous in 1911 was one of the first to demonstrate that viruses could cause tumours. Working with a transmissible tumour of chickens, he showed that cell free extracts of tumours (prepared by passage through filters capable of removing bacteria) would cause tumours — sarcomas — when injected into chickens. At about the same time Ellerman and Bang showed that chicken leukosis, a malignant disease of white blood cells, was also transmissible by bacteria-free cell filtrates. A considerable number of tumours in domestic and wild animals are now known to be caused by viruses (Table 5.1). In addition a number of viruses, which are not usually associated with tumours in natural conditions, can be shown to be carcinogenic under special circumstances in the laboratory. These viruses include members of the papova-, adeno- and herpesvirus groups (Table 5.2). Most of our knowledge about tumour induction by viruses has come from studies with papovaviruses and RNA tumour viruses (the group which includes the sarcoma and leukaemia viruses of Rous, Ellerman and Bang).

## 5.3 Papovaviruses

This virus group includes the wart (papilloma) viruses and polyoma virus and SV40. Polyoma virus was named thus by Stewart and Eddy, who showed that this virus — a harmless passenger in adult mice — caused a wide variety of tumours, including both carcinomas and sarcomas, in baby mice. SV40 (Simian Virus No. 40), an apparently harmless passenger virus in monkeys, also induces tumours in newborn

**Table 5.1** Spontaneously-occurring virus-induced tumours in animals

| Species | Tumour | Virus |
|---|---|---|
| Chicken | Leukaemia<br>Sarcoma | RNA tumour viruses |
| | Lymphoid tumours | Marek's disease virus<br>(a herpesvirus) |
| Mouse* | Leukaemia<br>Sarcoma<br>Mammary<br>carcinoma | RNA tumour viruses |
| Cat | Leukaemia<br>Sarcoma | RNA tumour viruses |
| Rabbit | Papilloma† (wart) | Shope papilloma virus<br>(a papovavirus) |
| | Fibroma† | Shope fibroma virus<br>(a poxvirus) |
| Monkey | Fibroma† | Yaba monkey virus<br>(a poxvirus) |

\* Inbred lines only
† Benign skin tumours; they usually regress but some rabbit papillomas become malignant.

**Table 5.2** Tumour induction by papovaviruses, adenoviruses and herpesviruses

| | Virus | Species | Tumour type |
|---|---|---|---|
| Papovaviruses | Polyoma | Mouse | Carcinomas of salivary gland, mammary gland, kidney, etc. |
| | SV 40 | Hamster<br>(newborn) | |
| Adenoviruses<br>(Human, simian<br>bovine and avian) | Serotypes vary<br>in oncogenic<br>potential | Rodents<br>(newborn) | |
| Herpesviruses | Herpesvirus<br>Saimiri (Squirrel<br>monkey herpes) | Marmosets<br>Owl monkeys | Lymphomas |
| | Guinea pig<br>herpesvirus | Guinea pigs | Leukaemia |

**Table 5.3** Properties of cells transformed by papovaviruses

Grow to higher densities*
Low amounts of serum required for growth†
Arrangement of cells in monolayers irregular†
Will grow in Agar†
Will form tumours in animals†
Agglutinable by plant lectins†
Antigenic properties of cell surface altered*
Rate of nutrient transport into cells increased*
Glycoproteins and glycolipids of cell surface altered*

\* Compared with normal cells
† Properties not shown, or shown to much lesser degree, by normal cells

animals – hamsters, in this instance.

All the papovaviruses have a similar structure (Table 2.1); much more is known about the virology of polyoma virus and SV40 as papilloma viruses are difficult to work with using currently available cell culture systems. The attractiveness of papovaviruses to the research worker stems, in part, from one very important fact – the smallness of their genomes. With only about 6000 base pairs the genomes of polyoma and SV40 can only code for polypeptides of 200 000 daltons total weight. The number of genes is therefore likely to be small and the prospect of determining their function – including those concerned with carcinogenesis – a realistic task.

### 5.3.1 Transformation by polyoma virus and SV40

Cells infected with polyoma and SV40 viruses can undergo one of two fates – they may support a lytic and productive infection, or they can support an incomplete, abortive infection. In the latter case, a minority of cells may undergo a change known as transformation. These cells show many of the properties characteristic of tumour cells. Whether a cell becomes transformed or supports a lytic infection depends largely on its genetic properties. For example, infection of mouse embryo cells with polyoma virus results in a lytic, productive infection, whereas infection of baby hamster kidney (BHK 21) cells with the same virus leads to the transformation of a proportion of the cells in the culture with the production of little or no virus. However, it is possible to transform cells which are permissive for lytic infection by infecting them with defective viruses – either produced deliberately by irradiation with UV light or contained naturally in stocks grown in the laboratory. This type of transformation presumably follows the infection of cells with a virus particle incapable of expressing the information required for lytic growth but with intact genes for the establishment and maintenance of transformation.

Several criteria have been used to recognize cells transformed after infection with polyoma virus and SV40. A non-selective method involves the plating out of infected cells at a low density with the subsequent differentiation of transformed and non-transformed colonies by morphological criteria – the rounded transformed cells grow in a more irregular manner and to higher densities than normal cells. The selective methods rely on (a) the ability of transformed cells to grow in gels of agar or Methocel (normal cells do not grow in these gels) and (b) the ability of transformed cells to form dense colonies in confluent monolayers of normal cells some 2–3 weeks after infection of the monolayer. These, and other properties of transformed cells are summarized in Table 5.3. It should be realized that transformed cells usually do not exhibit all these characters, and that the ability to produce tumours in animals may vary from one transformed cell line to another. The changes in growth patterns seen *in vitro* appear to be due to the release of transformed cells from control mechanisms responding to (a) contact with other cells (contact inhibition) and (b) depletion of growth factors in the serum present in tissue culture media.

Transformation by SV40 and polyoma virus is very inefficient, between $10^4$ and $10^5$ p.f.u. being required for each transformation event In many systems only a small proportion of cells in the culture become transformed, even after exposure to large amounts of virus; in the SV40-3T3 (mouse) cell system up to 40% of the cells can be transformed if very high multiplicities of virus ($10^6$ p.f.u./cell) are used. This system has been widely used to investigate the changes which occur on transformation. Initially, events similar to those occurring in the early phase of the lytic infection also occur in cultures containing cells which will be transformed. These include the increase in specific activity of a number of enzymes concerned with DNA synthesis (e.g. thymidine kinase, DNA polymerase) — it seems probable that this reflects increased synthesis of host-coded enzymes — and the induction of cell DNA synthesis. However, virus DNA synthesis and virus capsid protein synthesis do not occur, although viral RNA can be detected by hybridization techniques. As in lytically infected cells, a viral specific antigen — T antigen (T = tumour) appears in the nucleus of non-permissive cells a few hours after infection. In abortively transformed cells this antigen eventually disappears but it remains in those cells which have been stably transformed. T antigen is a protein whose appearance, as already indicated, does not require virus DNA synthesis. Immunological tests show that it is not related to any virus capsid proteins; it binds strongly to DNA. There is no direct evidence that it is virus coded, a situation similar to that found with the virus-specific transplantation antigens, which are found on the surface of tumour cells. These antigens are detected by a tumour rejection test which examines the ability of animals immunized by X-irradiated tumour cells or transformed cells to become resistant to subsequent challenge with transplantable tumours. The transplantation antigens induced by SV40 and polyoma virus are immunologically distinct, with no cross reaction; tumour rejection is mediated by immune lymphocytes.

At least one round of cell division is required before the transformed state becomes irreversibly fixed. The nature of this process is not known. It is clear, however, that many cells transformed by SV40 contain complete copies of the viral genome, although these cells are free of detectable virus when examined by standard techniques. When SV40 transformed cells are grown in contact with permissive cells, however, virus often appears in the culture after a few days. This process can be rendered more efficient by inducing the artificial fusion of transformed and permissive cells using killed Sendai virus. The molecular basis of this rescue phenomenon is not understood; usually only a minority of transformed cells produce virus on fusion with a permissive cell. Some SV40 transformed cell lines can be induced to produce virus on treatment with mitomycin C, proflavine or hydrogen peroxide. Another way of demonstrating complete viral genomes in transformed cells involves the extraction of DNA from virus free SV40 transformed cells. When this DNA is used to infect permissive cells, infectious virus is produced. Direct evidence of the presence of viral DNA in transformed cells has also been obtained by two types of nucleic acid hybridization techniques. In one method highly radioactive virus-specific RNA is prepared from virion DNA using *E. coli* RNA polymerase and is hybridized with cell DNA extracted from transformed or untransformed cells. The other method depends on the principle that the rate of reannealing of denatured DNA is proportional to its concentration – small amounts of highly radioactive denatured virion DNA are allowed to reanneal in the presence of large amounts of denatured DNA extracted from transform or untransformed cells. Both methods demonstrate that SV40 viral DNA sequences are present in transformed cells. Recent estimates put the number of viral genome equivalents per cell at about unity. The integration of viral DNA into the host genome was first demonstrated by Sambrook and co-workers in 1968, using the DNA/RNA hybridization technique. They showed that hybridizable SV40 viral DNA sequences remained associated with high molecular weight cell DNA after various denaturation procedures. Virus-specific RNA has also been found in SV40 and polyoma virus transformed cells by hybridization techniques. As expected from the lack of expression of late viral functions in transformed cells, the viral RNA sequences are largely of the early type. Large RNA molecules containing viral sequences are found in the nuclei of SV40 transformed cells; it is currently thought that the synthesis of these molecules, which are longer than transcripts of the viral genome, starts with the initiation of transcription at a cell promoter, with read-through into integrated viral DNA, or vice versa.

No evidence of infectious virus or infectious viral DNA has ever been found in cells transformed by polyoma virus or adenoviruses. This finding has been explained in the case of adenoviruses by nucleic acid hybridization experiments using defined segments of viral DNA prepared by specific cleavage with restriction endonucleases. These experiments have shown that lines of adenovirus type 2 transformed cells contain sequences homologous to only part of the viral DNA, in some cases only about 14% of the sequences in the whole genome being detectable.

The gene responsible for the transformed state have also come under attack from another approach – genetics. Despite considerable technical difficulties, a large number of mutants of SV40 and polyoma – largely *ts* and host range – have been isolated. Study of these mutants has significantly increased our knowledge of transformation by DNA viruses.

The SV40 *ts* A class of mutants are defective in a very early function of the virus which is required for both transformation and lytic growth. At high temperatures they induce the formation of T antigen and stimulate the replication of host cell DNA, but they fail to produce any viral DNA or capsid antigens and they do not transform non-permissive cells. Non-permissive cells can be transformed by such mutants at low temperatures, and temperature up-shift experiments show that the mutated gene function is required for the maintenance of transformation.

Recent work has shown that the *ts* A mutants map in the early region of the genome and current evidence suggests that this part of the SV40 genome codes for T antigen, the A gene product and the protein responsible for transformation, a strong possibility being that these are in fact one and the same protein. The other SV40 mutants (groups B, C and D) map in the late region and have abnormal virion proteins. Cells transformed by these mutants remain transformed at non-permissive temperatures and thus the products of the genes defined by these mutations are not necessary for the maintenance of the transformed state.

## 5.4 RNA tumour viruses
### 5.4.1 General properties
This is a very large group of similar viruses which often cause tumours – nearly always leukaemias or sarcomas – in many animal species (Table 5.1). Unlike the papovaviruses, many RNA tumour viruses rapidly transform all the cells in a culture with high efficiency, with a concomitant productive infection. Virus growth is not accompanied by cell death – a

feature which follows infection with members of some of the other RNA virus groups which mature by budding from cell surface membranes. Like these viruses, the RNA tumour viruses contain lipid and bud from the surfaces of infected cells. More is known about the avian tumour viruses than those of other species. This particular group of RNA tumour viruses can be further classified in two ways — into sarcoma and leukaemia (leukosis) viruses, and into groups arranged according to biological properties of the virus envelope. The first distinction — between leukaemia and sarcoma viruses — is not always clear-cut and in practice, those viruses which transform fibroblasts *in vitro* are termed sarcoma viruses and those which grow in these cells without causing transformation are called leukaemia viruses. Five envelope-determined sub-groups are recognized, each containing sarcoma and leukaemia viruses. The envelope of the virus contains macromolecules determining: (a) the host range of the virus; (b) whether sarcoma viruses will grow in or transform cells pre-infected with a leukaemia virus — if the two viruses are closely related, the sarcoma virus can not penetrate the leukaemia virus infected cell to initiate infection and vice versa; (c) the antigenic properties of the virus in neutralization tests.

Sarcoma viruses can be quantitatively assayed by scoring for foci of transformed cells on monolayers of fibroblasts. These foci are easily recognized as they contain large numbers of piled-up, rounded cells. The foci enlarge due to both the growth of transformed cells and the recruitment of neighbouring cells by infection with progeny virus. Leukaemia viruses are more difficult to assay as they produce neither transformed foci nor a cytopathic effect in the conventional way. Their growth can be detected and measured, however, by using immunological methods, by tests based on their interaction with sarcoma viruses (e.g. inter-

ference) and by assaying for reverse transcriptase. Under certain circumstances plaque assays are also available.

The RNA tumour virus particle has a complex structure (Table 2.1). Of particular interest are the nucleic acids contained therein. None of these are infectious when extracted from the virus particle but it is thought that a 60-70S RNA species is the viral genome. Many of the other nucleic acid components are thought to be derived from the host, including 4-5S transfer RNA-like molecules and a 7S DNA component. Their function is obscure. When the 60-70S RNA is denatured by heat or DMSO, it is reduced to 35S fragments which often can be resolved into two electrophoretic mobility classes. On the basis of this and other evidence two models of genome structure have been proposed. These are (1) a large genome made up of dissimilar segments whose sum of genetic information is equivalent to $10^7$ daltons of single-stranded RNA and (2) a small non-segmented polyploid genome made up of a collection of two or more identical or almost identical subunits each comprising only about $3 \times 10^6$ daltons of RNA. These models are the subject of current research, the second being favoured by most workers at present.

### 5.4.2 Growth and transformation
Little is known about the early stages of the RNA tumour virus growth cycle, apart from the role played by the virus envelope in determining the ability of the virion to penetrate different types of cell, and the electron microscopic observation that virus particles reach the nucleus soon after this event. Studies with inhibitors of DNA and RNA synthesis have played a crucial role in the study of subsequent events. Noteworthy is the fact that inhibitors of DNA synthesis, such as 5-fluorodeoxyuridine and cytosine arabinoside, block virus growth if they are present from the time of infection. That virus RNA-directed DNA

synthesis is required for growth has been suggested by studies using 5-bromodeoxyuridine. High multiplicity infections are more resistant to the photosensitizing effect of this compound (which is due to its incorporation into DNA) than infections at low multiplicity. Single cell clone studies with this inhibitor have also indicated that the virus does not stimulate DNA synthesis and that the effect of the inhibitor is not mediated through the killing of stimulated cells but rather by an effect on virus DNA. Actinomycin D inhibits the production of progeny virus when added at any time during the virus growth cycle, a finding which suggests that DNA-dependent RNA synthesis is required for growth. These observations, which sharply distinguish the RNA tumour viruses from other viruses with RNA genomes, were unified by Temin (1964) in his provirus hypothesis. He suggested that a double-stranded DNA copy (the provirus) of the RNA genome of the virus was synthesized after infection (explaining the sensitivity of this growth to inhibitors of DNA synthesis), and that further RNA synthesis was directed by this molecule — explaining the sensitivity to actinomycin D. Stably transformed cells could also be explained by this hypothesis by postulating integration of the provirus into the host genome — a situation analogous to transformation by papovaviruses. This bold hypothesis was not generally accepted and was shelved with unfavourable comments about the specificity of the metabolic inhibitors used — unwarranted as subsequent events showed. One of the requirements of the provirus hypothesis was an enzyme capable of synthesizing double-stranded DNA from an RNA template. This requirement was satisfied in 1970 by the discovery of such an enzyme — now called reverse transcriptase — by Baltimore and Temin and Mizutani in the virus particles of both avian and murine tumour viruses. This discovery aroused great interest — not only because the provirus hypothesis was now rendered respectable — but because the central dogma of molecular biology — DNA → RNA → protein — now required modification to DNA ⇌ RNA → protein.

### 5.4.3 Reverse transcriptase

Since its discovery, reverse transcriptase has been studied in great detail by many investigators. For biochemical studies the enzyme has usually been obtained from avian myeloblastosis virus (a leukaemia virus) because this virus occurs in large amounts in the blood of infected birds. This renders virus purification a relatively simple task. The enzyme may be readily assayed after treating virus particles with non-ionic detergents. Under these conditions deoxynucleoside triphosphates are incorporated into DNA. This DNA can be hybridized with 60-70S virion RNA, indicating that this is the template for its synthesis. The enzyme can be relatively easily purified as it is released from the virion in soluble form by non-ionic detergent treatment, a feature which strongly contrasts with the behaviour of transcriptases from other RNA and DNA viruses. Of considerable help during purification procedures is the ability of the enzyme to copy templates other than virion RNA. The amounts of DNA synthesized in the presence of these polymers (which include poly (dA-dT) poly (C) and poly (A)) is 10 to 1000 times that made in the presence of virion RNA alone. Like other DNA polymerases the reverse transcriptase is primer dependent — it cannot initiate new chains of DNA but only extends preformed chains. For example, poly (A) is a very poor template alone, but in the presence of a small amount of oligo (dT), extensive poly (dT) synthesis occurs. As 60-70S virion RNA without added primer is a relatively good template for the enzyme it seems likely that it contains primer; present evidence indicates that this is a small RNA molecule which can be removed from the 60-70S RNA by heating.

These primers probably exist at a large number of sites on the 60-70S RNA, as the DNA molecules made are all very short (about 100 nucleotides) but are complementary to all regions of the RNA template. It should be noted, however, that most of the DNA made is complementary to only a small portion of the template. A considerable proportion of the DNA molecules made are double stranded, and an enzyme, ribonuclease H, has been found intimately associated with reverse transcriptase. It has been shown that this enzyme performs the functions necessary for the initiation of synthesis of complementary DNA chains as from a DNA-RNA hybrid it removes the original RNA template and also leaves a small RNA primer. Although much is known about the *in vitro* properties of reverse transcriptase its role during infection and transformation is still poorly understood. Studies with two types of mutant of Rous sarcoma virus indicate that the enzyme is essential for virus growth. Hanafusa and Hanafusa have described a non-infectious mutant (RSV $\alpha$) which cannot infect cells even when helper virus is present or when blocks to penetration are circumvented; this virus contains 60-70S RNA but lacks reverse transcriptase. Temperature sensitive mutants with temperature sensitive reverse transcriptase activities have also been isolated; these viruses are able to penetrate cells at non-permissive temperatures but they are unable to grow or transform cells under these conditions. On the other hand, these viruses can initiate infection if kept at the permissive temperature for some hours after infection, but then become resistant to a shift-up in temperature, suggesting that the temperature sensitive function is only required for the initiation of infection and is not needed for the subsequent maintenance of virus growth and transformation.

Direct evidence for the existence of a DNA provirus has been provided by nucleic acid hybridization studies and by the isolation of infectious DNA from transformed cells. Hybridization studies have been rendered difficult to interpret by many problems, including the presence of host RNA species in virions and the universal presence in chick cells of the provirus of avian leukaemia viruses. However, recent experiments using mammalian cells transformed by Rous sarcoma virus have shown that these cells contain 1–2 genome equivalents of viral DNA. Before infection these cells contain no detectable viral DNA. Carefully controlled experiments by Hill and Hillova have shown that infectious DNA can be isolated from mammalian cells transformed by Rous sarcoma virus, providing convincing evidence that the provirus exists and that it contains the genetic information required for the production of infectious virus particles.

### 5.4.4 Genetics of RNA tumour viruses

Mutants of avian RNA tumour viruses can be divided into two classes — non conditional and conditional. Non-conditional mutants have been known for many years, the best known example being the Bryan high titre strain of Rous sarcoma virus (BH.RSV). This virus can transform cells, but produces non-infectious particles unless a helper virus is present. Stocks of BH. RSV in fact contain a helper virus; this is called Rous associated virus (RAV) and is a leukaemia virus. This situation, worked out by Temin, Vogt and Rubin, explains why infections initiated by single virus particles (i.e. at high dilutions) lead to transformation but not virus growth, whereas high titre virus is produced by infections initiated by large numbers of particles (i.e. at low dilutions). The latter conditions are of course required for simultaneous infection of cells with both BH.RSV and RAV. The defect of BH.RSV lies in a surface glycoprotein and this defect is complemented by phenotypic mixing with the envelope glycoprotein of the RAV. The infectious BH.RSV now shows the envelope properties of the RAV (e.g. host range) and its pseudotype is indicated as follows: BH.RSV (RAV –2), the number indicating the helper virus type. Recombination does not appear to occur during mixed infection with defective sarcoma viruses and their helper viruses, although it occurs freely between non-defective sarcoma viruses (i.e. helper independent) and leukaemia viruses. Host range and trans-formation markers were used in these experi-ments. Recombination studies have only recently started with RNA tumour viruses and more work is needed to explain these phenomena. Other classes of non-conditional mutants include a group of transformation-defective sarcoma viruses. These arise spon-taneously at high frequencies in stocks of non-defective sarcoma viruses. They retain the envelope properties of the parent virus and are therefore unlikely to be recombinants between the parent virus and helper viruses present in the cells used for virus propagation. Bio-chemical evidence suggests that these viruses may lack the larger of the two electrophoretic mobility classes of genome fragment produced by heating the 60–70S viral RNA. They thus resemble leukaemia viruses which lack this RNA, and can probably be considered to be deletion mutants. These mutants show that sarcoma virus transformation and growth can be clearly separated, the former not being required for the latter. Another class of non-conditional mutant includes viruses which give rise to transformed cells with abnormal morphology. Commonest among these are those which induce fusiform (spindle shaped) transformed cells, rather than the usual type of rounded cell.

A considerable number of avian sarcoma virus *ts* mutants has been isolated usually after mutagen treatment. These fall into three physiological classes: (1) those which grow but do not transform at the non-permissive tem-perature; (2) those which are temperature sensitive in both growth and transformation; and (3) those which transform but do not grow at the non-permissive temperature. The majority of mutants isolated fall into category (1). They clearly show that transformation is under the continuous control of virus genes. Among the *ts* properties of cells infected with these mutants include transformation-associated characteristics such as morphological changes, an increased rate of sugar uptake and an increased susceptibility to agglutination by plant lectins. Important experiments where mutant-infected cells have been transferred from non-permissive to permissive temperatures have shown that transformation can occur rapidly on release of the *ts* block (50% transfor-mation within 6h.), this requiring protein synthesis but not RNA or DNA synthesis. Transformed cells become like normal fibro-blasts equally rapidly on temperature shift-up. One mutant in this class causes infected cells to become transformed after temperature shift-down within one hour; protein synthesis is not required for this. Evidence from complemen-tation tests indicates that class (1) contains mutants defective in different genes. Class (2) contains the *ts* reverse transcriptase mutants referred to above. Genetic and biochemical studies on these mutants are still in their early stages but we may confidently predict that they will provide us with much more useful infor-mation about the molecular basis of transfor-mation.

## 5.5 Comparison of transformation by DNA and RNA tumour viruses; oncogene theory

DNA and RNA tumour virus mediated cell transformations possess many features in common. Noteworthy is the continuous requirement of a virus gene product(s) for maintenance of the transformed state. The synthesis of this product(s), as yet unidentified, is directed by viral genes which are almost certainly integrated in the host chromosomes. The continuous maintenance of the transformed state by the virus sharply distinguishes virus carcinogenesis from tumour induction by agents such as chemicals and radiation, which are only required for the initial steps of tumour formation and are not needed during subsequent tumour growth. This difference may be more apparent than real, as suggested by the oncogene theory of Huebner and Todaro. This theory, which has not yet been critically tested, proposes that all cells contain the genetic information necessary to specify the genome of RNA tumour viruses (the virogene) and the genome containing transformation genes (the oncogene). The expression of this information is normally repressed, but under certain conditions (e.g. treatment with chemical carcinogens or irradiation), some, or all, of it is expressed with either tumour formation or virus production or both, depending on which parts of the virogene are derepressed.

# Suggestions for further reading

Molecular virology is a rapidly advancing subject and textbooks become rapidly out-of-date. We have therefore selected three categories of further reading – works which will introduce the reader to the basic principles of molecular biology, up-to-date monographs on virus groups, and periodically appearing reviews.

## Introductory works

Watson, J.D. (1970) *Molecular Biology of the Gene*, 2nd edition, W.J. Benjamin, Inc., New York.
  A clear and comprehensive introduction to molecular biology.
Hayes, W. (1968), *The Genetics of Bacteria and their Viruses*, 2nd edition. Blackwell Scientific Publications, Oxford.
  The standard text.
Stent, G.S. (1971), *Molecular Genetics*. W.H. Freeman & Co., San Francisco.
  An excellent general introduction.
Cairns, J., Stent, G.S. and Watson, J.D. (eds) (1966), *Phage and the Origins of Molecular Biology*. Cold Spring Harbor Laboratory.
  A collection of essays by phage workers and animal virologists associated with M. Delbrück. Many classical experiments described in autobiographical fashion by the scientists involved.

## Monographs, etc.

Fenner, F., McAuslan, B.R., Mims, C.A., Sambrook, J. and White, D.O. (1974), *The Biology of Animal Viruses*, 2nd edition. Academic Press, London, New York, San Francisco.
  The standard text – has no rival.
Tooze, J. (ed) (1973), *The Molecular Biology of Tumour Viruses*. Cold Spring Harbor Laboratory.
  The only up-to-date and comprehensive review. Authoritative.
Hershey, A.D. (ed) (1971), *The Bacteriophage Lambda*. Cold Spring Harbor Laboratory.
  The first fifteen chapters are essential reading for anyone wishing to get to grips with this fascinating and important virus.
Dalton, A.J. and Hagenau, F. (1973), *Ultrastructure of Animal Viruses and Bacteriophages: An Atlas*. Academic Press, London, New York, San Francisco.
  An authoritative and comprehensive account of virus ultrastructure – copiously illustrated.

## Reviews

*Advances in Virus Research*. Published annually by Academic Press. Contains authoritative reviews on viruses of plants, animals and bacteria.
*Annual Reviews of Microbiology*. Published by Annual Reviews Inc., Palo Alto. Each volume usually contains several comprehensive reviews on virological topics.

# Index